MANAGING
SNOW&ICE

SECOND EDITION

BY JOHN A. ALLIN

EDITOR
Jeff Fenner

GRAPHIC DESIGNER
Nichole Frye

PRODUCTION MANAGER
Helen Duerr O'Halloran

COPY EDITOR
Lisa Rose

Publisher: Kevin Gilbride
GIE Media, Inc.
© Copyright 2011

Address all correspondence to GIE Media, Inc. at www.giemedia.com or call 800/456-0707.

Library of Congress Control Number: 2011926561
ISBN: 978-1-883751-33-3

www.snowmagazineonline.com

CONTENTS

ABOUT THE AUTHOR

John Allin is a full-time consultant catering to contractors in the snow and ice management industry. He has been involved in the snow industry since age 16, starting out plowing snow for his father's business in New Jersey while still in high school. As part of his consulting practice, Allin also works with those in the legal arena as an expert witness.

Prior to establishing his consulting practice in 2008, Allin was president and part-owner of Snow Dragon Snowmelters. This patented, portable melting process came about as a result of the 2002 Winter Olympic Games in Salt Lake City, Utah, where Allin's company – Snow Management Group – managed all mechanized snow removal activity for the games.

Snow Management Group was originally established in 1978 and eventually managed snow on more than 5,000 sites in 42 states. The company was recognized as the first national snow contracting business. Allin sold the company in 2004.

He is considered the foremost snow industry expert in the world and has consulted on snow management projects in Europe, China, Scandinavia, Australia and Russia. Some of the projects he has consulted on include the City of Bejing; Newark International Airport; JFK International Airport; the 2010 Winter Olympic Games in Vancouver, BC, Canada; Lucent Technologies; the east coast distribution center for The Gap; Falls Creek Resort in Victoria, Australia; and the City of Moscow. Allin also set up worldwide distribution for the Snow Dragon Snowmelters, having secured and visited 42 distributors in 36 countries.

Allin is one of the founders of the Snow & Ice Management Association (SIMA) and served as board president from 1996 to June 2002. In 2010, GIE Media's *Snow Magazine* honored Allin with a Pioneer Award for his service to the snow industry. Allin and his wife, Peggy, live in Erie, Pa.

ACKNOWLEDGEMENTS

In some ways, authoring a second edition of a book is easier and harder at the same time. I had to read every chapter, word for word, determining what was "old" and should be discarded. At the same time, I had to consider the new things that should be included in a manuscript of this nature. Seeking and obtaining guidance from others is paramount to the success of such an endeavor.

For the first book, I was blessed with an editor who understood the process and was incredibly helpful to me. He made me look good. For this book, lightning struck twice, which I'm told is highly unusual. I was concerned about the process and how I would interact with someone "new." Suffice it to say, Jeff Fenner was the consummate professional who fully understood the challenges associated with what I was going through as I churned through the writing process. He guided me as one would nurture an anxious child out on his own for the first time. For that I am grateful. He made it "easy" to work through the process and, of course, made me look good.

There are few people in this world who I consider "friends." In fact, I can count them on one hand. Most people I interact with are acquaintances of varying degrees, some closer than others. One person was there as a sounding board through the ups and downs of the past seven years of my career. Not everything we do turns out as planned or as expected, but true friends don't judge. They listen and offer consolation, encouragement or advice. There have been a few people in this world who have impacted me as a person and as a professional. Kevin Gilbride is one of those very few. He is also the publisher of this book, and his counsel has always been well grounded. I have been truly blessed to have been able to call him a "friend."

Charles Glossop had nothing to do with the writing of this book, but he is one of those other "few" who I consider a true friend. Together we have celebrated our various successes in our lives, and we have commiserated during the low points. We talk regularly about how our mothers are fairing as they both move through the autumn of their lives. We also share the common bond of having been there at SIMA's birth and both care deeply about the snow industry. At some of the lowest points in my life, he was there and never judged me. I don't think we have ever had a disagreement, and I owe him an incredible debt of gratitude for just "being there."

My father, Whitey Allin, continues to be someone I can lean upon whenever necessary. His guidance and advice have allowed me to venture into uncharted waters, knowing I am not alone in my quest. I shall be forever grateful for having had him in my life. I hope and pray that continues to be the case for many, many years to come.

John A. Allin
June 2011

DEDICATION

Behind every successful man is a good woman, so the saying goes, and it is certainly true in my case. Smarter than I am, yet patient enough to go along with some of my more harebrained ideas, my wife is the center of my universe – even though she does not think so.

When I wanted to do the 2002 Winter Olympics in Salt Lake City, she thought I was out of my mind. But when it became apparent I was going to go for it, she jumped on board and helped me with a dedication unmatched by anyone I know. I do not disagree that I am difficult to live with. I am also headstrong to a fault. When I was traveling around the world chasing my dream of taking a new, untested product to Europe, Asia, Russia, Australia, New Zealand and Scandinavia, she was there when I came home and encouraged me to continue upon the course of action I found myself on.

When I was deciding what to do after Snow Dragon, she pushed me to begin yet another career with my consulting practice. I wonder if it was because she grew weary of me hanging around the house not accomplishing much, but I know it was because she believed in me and knew I could do it, and do it well.

Nothing I have achieved in the last 25 years would have been possible without her love and support. For that I am a very lucky man. I pray to God that when it comes time for one of us to depart this earth, I go first. I don't want to go through what's left of this lifetime without her by my side. She is my partner in life's journey. I love you, Peggy, and don't know what I would do without you.

John A. Allin
June 2011

Chapter One
A HISTORY OF SNOW MANAGEMENT

Chapter Highlights

- Early Snow Management
- The Modernization Of Snow Management
- An Industry Evolves

As families across the city made their final preparations for the 2001 Christmas holiday, snowflakes descended from the skies over Buffalo, N.Y. Falling heavier by the minute, it became apparent this was no ordinary snowstorm. More than 2 feet of snow hammered Buffalo on Dec. 24, effectively shutting down the city and Buffalo Niagara International Airport.

Three days later, another 25.6 inches of lake-effect snow piled up, creating a total of more than 4 feet between Dec. 24 and 27. The 29.8 inches of snow that fell from Wednesday, Dec. 26 through Thursday, Dec. 27, was the second snowiest 24-hour period in Buffalo meteorological history.

In total, nearly 7 feet of lake-effect snow fell on the city between Christmas Eve and New Year's Day, prompting then President George W. Bush to declare a state of emergency for Buffalo and upstate New York.

The storm was brutal, causing many injuries and countless economic losses due to temporary business closings. But the storm's damaging impact would have been much worse, if not for the efforts of snow and ice management professionals, who helped dig out the city inch by inch. After a few days of intense work, life slowly returned to normal in Buffalo.

Stories like this, although on a smaller scale, happen every winter across much of the northern half of the U.S. and throughout Canada. Snow and ice professionals work under harsh conditions and for long hours to fight Mother Nature in her own backyard. Commerce, transportation and life as we know it would not proceed in the winter without these dedicated professionals.

The work isn't completed out of sheer selflessness, however. Snow and ice removal, if managed properly, can be very profitable. Indeed, a few in the industry make their entire year's revenue from just a few months of snow and ice management work.

Early Snow Management

Those dedicated snow and ice professionals across North America are part of a long line of individuals who, since the modern age began, have been finding ways to deal with winter precipitation. According to "Have Snow Shovel, Will Travel" (National Snow and Ice Data Center, www.nsidc.org), the earliest attempts at snow control in North America involved leveling snow drifts for sleigh traffic during the 1700s. At the time, many city laws required business owners to clear snow from the sidewalks and streets in their area, and wintertime travel in the early 1800s

Responding to extreme snow events is a part of daily life in snow and ice management. Note the cars buried in snow.

was mostly done on foot.

The 19th century brought increased industrialization and commerce to North America, making snow management more important. In the 1840s, inventors designed and patented the first snowplows, but several years passed before the plows were widely used. Reportedly, one of the first mentions of snowplow use comes from Milwaukee in 1862, where a plow was attached to a cart pulled by a team of horses. Throughout the next several years, horse-drawn plows gained popularity and came into use in many other northern cities.

Soon after plows were implemented on a wide scale, retail businessmen began complaining, claiming that their storefronts were completely blocked with mounds of snow, making them inaccessible to customers. Sleigh drivers also found fault with the plowing system because of the ruts and uneven surfaces it created.

Cities along the eastern U.S. and in the Snowbelt responded by hiring horse-drawn carts and individuals with shovels to work in conjunction with the plows, hauling away the plowed snow and dumping it into rivers – a definite "no-no" in today's environmentally conscious world.

The Big One. A major push for advancement in snow management techniques came as a result of the notorious "Blizzard of 1888," which paralyzed the Northeast after three days of snow, wind and freezing temperatures. Reportedly, snow covered the entire first levels of buildings, as more than 2 to 4 feet of snow fell and drifted.

The storm took the lives of more than 400 people and immobilized fire departments, passenger trains and communication across much of the Northeast. Communication was cut between New York, Boston, Philadelphia and Washington, D.C., for several days when telegraph and telephone wires were downed by high winds and heavy accumulations of ice and snow.

Following the 1888 blizzard, cities recognized the need for more organized snow removal practices. They were motivated to bury communication wires and create alternative methods of transportation, such as trains and subways, which wouldn't be hindered by the accumulation of drifting snow.

Here Come The Cars. Once the automobile became popular, an increasing demand was placed on removing snow from city streets. Likewise, cities began "motorizing" their snowplow fleets using dump trucks, tractors and other heavy equipment for snow removal.

In 1920, the Barber-Green snow loader was invented and successfully introduced in Chicago, and several cities purchased the machine thereafter. The snow loader used tractor treads, a scoop and a conveyor belt. As the snow was plowed, it was forced up the scoop, caught by the conveyor belt and then carried up and away from the street into a chute at the top. The snow was dropped into a dump truck that was parked underneath.

Around the same time of the Barber-Green snow loader, financial pressure from the loss of business due to snow and ice storms prompted cities to invest in new fleets of dump trucks and tractor plows.

Ice left behind by snowplowing called for increased use of salt and sand. With the need for clear streets outweighing environmental concerns, city public works officials used salt by the ton to ease road conditions, while also experimenting with cinders and sand.

Truck-mounted salt spreaders became the primary tool in fighting snowy roads. Businesses and private citizens used tons of salt to keep driveways, sidewalks and access routes clear of snow and ice.

Snow Management's Modernization. Advances in weather forecasting, including the use of satellites, began to assist snow professionals in the late 1950s and 1960s. As a result, snowplow operators could prepare in advance for major snow and ice storms.

As equipment became more efficient and effective, plow technology advanced, as well. Hydraulics, introduced in the 1960s, allowed for more efficient plow operation. They made it easier to maneuver the plow setup and allowed operators to stay inside the cab to move the plow from side to side.

The addition of electronic switches enhanced operations, but the mechanics of moving snow from pavement surfaces remained basically the same. Loaders made it easier to move large quantities of snow, but were very inefficient since the operator was using a large piece of equipment to move a relatively light amount of snow. Skid steers have made sidewalk snow clear-

> "Unfortunately, **plowing snow** is done today much like it was decades ago."

ing quicker, as these units are considerably more productive than a human operating by himself or herself.

Unfortunately, plowing snow is done today much like it was decades ago – we push snow out of the way, usually to the left or right of vehicles. Advances in the technology of actually moving snow have been minor and mostly limited to making the ease of operation more attractive. The advent of motorized vehicles made the process faster, but the process is virtually still the same.

An Industry Is Born

Long after automobile use became widespread, shopping centers, office parks and industrial centers realized the need to have their parking lots and pedestrian areas free of snow and ice. Many of these companies turned to private snow management contractors or purchased their own equipment.

Today, the private snow management industry is a growing and well-established market. In 1996, the Snow & Ice Management Association (SIMA) was formed over the course of a June weekend by eight dedicated individuals. The founders hammered out the details for starting the association with little more than a desire to increase the level of professionalism among those who move snow for a profit.

Until this time, the snow "industry" was fragmented and unsophisticated. Various local companies trying to make a little extra money during the winter made up the majority of the private, commercial snow clearing "industry." While numerous state Departments of Transportation (DOT) poured money into research and development of equipment and materials to fight snow, the private contracting business had no dominant players on a national level, but strong local and regional contractors.

Landscape contractors, excavation contractors and other seasonal-type businesses did snow management to generate revenues during the slower winter months. However, in the mid-1990s there were no dedicated "snow-only" contractors. The formation of SIMA marked the beginning of a process whereby snow contractors could learn standard business practices.

This author was heavily involved in the growth of the snow contracting industry. As one method of growing SIMA's membership, I offered to speak on snow contracting at various industry trade shows. The former Associated Landscape Contractors of America

(ALCA) – now known as PLANET – knew they had members who were in the snow removal business in the winter months, and also realized they were not providing educational resources for this add-on service. ALCA embraced the birth of SIMA and allowed it to provide educational and practical support to its membership. The landscape association encouraged its members to join SIMA and provided considerable moral support to the leadership of the fledgling association of snow professionals.

While SIMA considers itself the voice of the snow industry, it is hard to accept that an association with 1,500 members speaks for an industry which today boasts upwards of 40,000 snow and ice contracting entities. However, there is little doubt SIMA has been instrumental in the sophistication of the industry at large. The industry now supports two trade magazines dedicated to the private snow contracting industry. The educational material available to industry professionals is head and shoulders above what was available back in 1996.

Though no hard numbers exist, it is believed that roughly 40,000 private companies across the U.S. perform snow and ice management, generating approximately $6 billion in annual revenue. That's not to mention the countless numbers of subcontractors and "one-man" operations who also remove snow and ice.

In 2001, successful plowing contractors throughout the U.S. and Canada indicated that gross profit margins for snow management exceeding 60% were normal. And gross profit margins for ice control services in excess of 70% were achievable. Those contractors who viewed snow management as

Before development of modern snow-fighting techniques, winter storms could paralyze travel and communication for long periods of time.

a profit center regularly stated that it was the most profitable part of their businesses. From contractors who project five snow events per season in a great winter, to those who project 35 snow events in a mild winter, in the 1990s and the early part of the 2000s, all reported gross profits at or in excess of those mentioned earlier.

An Industry Evolves

In 2008, the U.S. entered the early stages of a severe recession. Snow contractors were initially spared, but by 2010 even those in the snow industry felt the effects.

Margins were squeezed as customer purchasing departments were exerting downward price pressures. Snow contractors were not spared. Established contractors felt the squeeze as new competition entered the market. Individuals laid off from full-time jobs put plows on their four-wheel drive vehicles in unprecedented numbers and attempted to make extra money plowing for homeowners and companies.

While some elected to become subcontractors for established snow entities, others became individual operators, securing business by cutting quoted prices to break into the market. Purchasing departments challenged contractors to produce the same quality service at lower prices, and this led to the lowering of generally accepted margins across the country. Margins fell from the mid-60% range into the "new norm" of upper 40% gross margins. As a result, contractors learned to do more with less.

In the 1990s, customers who paid $25 for a selected service found they were able to pay $18 for the same service in 2010. Those "in-the-know" predict the trend could continue for several years before any sort of rebound occurs. And even then, any such "rebound" will likely be modest at best. The days of 60% and 70% across-the-board gross margins are likely gone forever.

However, as with any recession, the enterprising contractor who weathers the storm will likely come out leaner, stronger and better able to do business under the new normal rules of business engagement.

17

Chapter Two
GETTING STARTED

Chapter Highlights

- Who Are Snow Management Professionals?
- Things To Consider
- What You Need To Get Started
- Working As A Subcontractor
- Finding The Right Partner
- Snow Management As An Add-On Service

Entering the snow management business once required only dedication and perseverance. As with any business, a snow and ice management company needs to be managed effectively and profitably. Ten years ago, when presenting to a room of 100 snow contractors, it was normal to find four or five company owners with college degrees, and none would have advanced degrees in accounting or business management. Fast-forward to today and there will be 35 to 50 people in the room who have college educations, many with advanced degrees.

This is evidence that the sophistication level of snow and ice industry has advanced considerably. In the not-so-distant past, snow management was considered an ancillary service to another core business entity. These core businesses included landscape contracting and maintenance, pavement maintenance, excavation and/or trucking. The decision to enter the snow management business for these entities was driven by the need to increase cash flow during slower times of the year or by customers who demanded the service.

Today, the decision to enter the snow and ice management business is more often dictated by profit potential. As a result, the number of companies exiting their core businesses and entering snow and ice management full-time is on the rise. This influx of new companies calls attention to the fundamentals of how to run a snow business full-time, year-round.

In the industry's infancy, large snow companies operated in tight, well-defined geographic markets. The number of dedicated snow companies grew as the demand increased. Today, there are dozens of companies – several of which can be considered "national" – that are doing snow and ice management exclusively.

To give you perspective on the industry's growth, consider that 10 years ago the largest snow contracting company in North America generated $28 million in revenue. In 2010, the top grossing company was well over $100 million.

That company, the Brickman Group, is well-capitalized and structured to support growth and expansion. Interestingly, this company is not a "snow only" organization. Brickman relies on branch

The professional snow management industry has grown significantly over the last decade.

operations strategically placed around the U.S., with a core competency based in landscape maintenance. The company has a stellar reputation for honesty, and it treats its employees and subcontractors fairly.

To further demonstrate the industry's growth, 10 years ago there were eight companies with a $1 million or more in revenue. By 2010, there were 96 companies with gross revenues in excess of $1 million. (Source: *Snow Magazine*)

Previously, the desire for additional cash flow or keeping your workforce busy in the "offseason" were valid reasons to enter the snow and ice management industry. Of course, that may still be true in some instances, but a more appropriate reason is to develop a profit center that will augment the existing core business.

The "old rules" no longer apply. To be successful in the snow and ice management industry, one needs to understand financial goals, gross margin potential, capital reserve requirements, equipment procurement issues and the impact of proper human resource management. Sophistication is a double-edged sword. Without some sort of business acumen, those who enter the business for a "little extra money" rarely survive past the third winter.

Who Are Snow Management Professionals?

Landscape contractors have a natural avenue into the snow business since their customers often require winter services. Such contractors have a good portion of the equipment already available, and the jump from two-wheel-drive vehicles to four-wheel-drive vehicles is not difficult. Further, the landscape contractor also has ready access to the personnel required for snow operations and has the mechanical ability to fix inevitable equipment breakdowns.

Pavement installation and maintenance contractors, as well as excavation contractors, who have access to larger equipment, are often involved in larger commercial plowing accounts. When a melding of these contractors happens, the ability to increase service capability and capacity is possible. Thus, the very largest of accounts can be quoted and serviced.

Contractors considered "snow-only" often bring all of the previously mentioned contractors to the party. These contractors think snow year round, embrace the latest technology and institute the latest productivity techniques. They research new market trends, and often are instrumental in inventing or contributing to new technology.

In the past 10 years, the lines between snow equipment manufacturers and those who use the equipment have blurred. Some very progressive snow contractors now design new equipment. Likewise, recent advances in software technology have come from the minds of well-established, progressive snow contractors, not outside programmers. Technology advances will be discussed in greater detail throughout this book.

Things To Consider

No matter your business model or motivation for entering into the snow management industry is, there are several questions to consider before getting started.

- How much money do you invest?
- Do you use your own equipment, or do you use subcontractors?
- In growing this portion of your business, do you concentrate on the residential or commercial market – or both?

There are several important issues you must consider to become a full-time, year-round snow contractor and make it your core business. First, your business cannot be operated as a temporary or secondary venture. If you are going in, then put both feet through the door. The second, you must institute formal accounting procedures, including proper capitalization and cash flow planning. Finally, a sales plan is needed to secure customers and start the revenue flowing.

High Stress. Snow is usually a high-margin business, even with the downward price pressures the industry has experienced in recent years. However, that money can be clouded by sleep deprivation, angry spouses, disappointed children and unfeeling customers. Before you start, make sure you have seriously considered your level of commitment. Even for the year-round, snow-only contractor, the in-season stress levels can extend into the summer months. Continual marketing of the service offering, budget reviews, hiring and sales presentations, often to individuals and companies who don't want to think about snow when the outside temperatures are pushing 80 degrees, will be part of the to-do list.

Demanding Customers. Snowplowing isn't for everyone. It requires long hours, creates sleepless nights, and introduces you to demanding customers who aren't interested in how much – or how little – sleep you get. They want their snow and ice gone. Snow falls at very inopportune times. Missing a family Hanukkah, Christmas, New Year's Eve – or even Thanksgiving – is a very real possibility. Customers will often call just as the first gift is being opened because they need to get to family gatherings, parties, or to the store for that last-minute item.

Unfortunately for the snow contractor, customers expect a high level of service even when the average price for

SNOW FACT:

In the early 1900s, skiers created their own terminology to describe types of snow, including the terms "fluffy snow," "powder snow," and "sticky snow." Later, the terminology expanded to include descriptive terms such as "champagne powder," "corduroy," and "mashed potatoes."

your service goes down. Standing one's ground when pricing services is essential to sustain viable profits and helps you build the capital required for future growth. It takes intestinal fortitude and commitment to stay true to your business plan.

For example, a residential customer once called me at 8 a.m. on Christmas Day demanding we plow his driveway so he could attend church services. Sixteen inches of snow were already on the ground and city plows hadn't been dispatched. I plowed the driveway and explained to the homeowner that he would only be able to make it to the end of the driveway.

After plowing out another customer with a similar intention of going out to get a loaf of bread, I went by the first customer only to see his car stuck at the entrance to the driveway. I got no satisfaction from being correct.

Another example of a demanding customer is the property manager who oversaw a retail site with 45 acres of pavement. He questioned why we plowed his site when "very little snow" had fallen. The manager, who lived 15 miles from the site, posed this question since there was only a dusting of snow on his driveway. He felt the need to participate in the decision-making process from that point forward, and we agreed to do so to keep the account.

Several days later I was at the site at 2 a.m. and found 4 inches on the pavement with snow continuing to fall. I called the manager at home to explain the situation and asked "permission" to plow the lot. He agreed. The next night the same thing occurred, and I called him again, this time at 3 a.m. His wife answered the phone and I cheerfully asked for our contact, explaining I was at the site and needed to talk to him.

This went on for two more nights. After the fourth early morning wake-up call, he decided we could make the determination ourselves going forward.

This attitude, especially on the part of the residential customer, is common, frustrating and often difficult to envision before you enter the business. However, this comes with the package when you choose to enter the snow and ice management business.

Commercial customers are no less demanding. Retail facilities are now open on most holidays and often demand their parking lots and facilities be free of snow 24 hours a day, seven days a week. Regardless of your own personal feelings, the customer comes first in such situations if you hope to achieve the customer satisfaction levels necessary for your business to succeed and grow.

It is unfortunate that some consumers who are unfamiliar with the service we provide, the dedication it requires, and the financial commitment needed, believe those who plow snow "cannot find real jobs." This is not the case with the majority of customers, but it is a stereotype anyone who enters this business must deal with on a regular basis.

What You Need To Get Started

Some people enter the snow business with a snow blower and a snow shovel. This is the business at its most basic level. Usually, this person has yet to acquire a driver's license and may be in high school or college. Though simple, this model is profitable if addressed properly. For example, a contractor in Massachusetts started out at the age of 9 shoveling residential walks and driveways in his neighborhood. Fast-forward 30 years and his company has become a nationally recognized snow and ice management operation.

Snow blowers cost anywhere from $350 to $1,500 depending upon the specifications (*See Chapter 11, Snow Management Equipment, page 107*). Yet, it is difficult to grow the business to any significant level if you only have a snow blower or shovel at your disposal.

Vehicles. Growing your business requires a basic plowing vehicle, usually a basic four-wheel-drive pick-up truck mounted with a snowplow. All the major truck manufacturers have their good and bad points in relation to snow removal. I have owned most types and do not necessarily favor one over another. Specific manufacturers aside, there are some things you should look

came quite reliable.

As with all new things, the initial pricing for "V" units was high, making it difficult for industry "newbies" to afford them. However, competition lowered the pricing and allowed most contractors to take advantage of the "V" blade's increased productivity. In the early 2000s, the Blizzard plow entered the market with an advanced design featuring hydraulically operated extensions on either end of the moldboard. After correcting flaws in the original design, the plow has become a popular choice with snow professionals looking to increase their productivity. Consolidation among plow manufacturers has also brought the price down on this plow type, allowing start-up companies to enjoy increased productivity.

"**Heavy-duty** – or contractor-grade – plows cost as much as 20% more than medium-duty plows."

for in selecting and ordering a pick-up truck for plowing. Four-wheel-drive trucks are usually recommended for plowing snow. Two-wheel-drive trucks can work if tire chains are used or if the truck is large enough to hold the required weight to allow the drive wheels to have traction (*See Chapter 11, Snow Management Equipment, page 107*).

Plows. The next investment is acquiring a plow for your truck. Over the past 10 years, the options for truck-mounted snowplows have grown considerably. In the early 1990s, the only real choice was what length blade to buy. There were 7½- and 8-foot blades, that's it. The original "V" blades were an engineering disaster, but once the bugs were eliminated these blades be-

Numerous manufacturers offer straight-blade plows, which come in several "duties" depending on the application. Light-duty plows are for smaller vehicles (F-150-size units), generally used for residential plowing operations. Generally, medium-duty plows are attached to new trucks at dealerships. These plows are reliable for residential plowing but can wear down fairly quickly when used for commercial plowing requiring multiple visits. Hydraulic lines can fail and structurally these units are not considered "contractor-grade."

Heavy-duty – or contractor-grade – plows cost as much as 20% more than medium-duty plows. "V" blades and blades with hydraulically operated "wings" are built to be production

A straight-blade plow mounted on a pickup truck is by far the most common snow removal tool for residential and light commercial work.

plows as they feature oversized fittings, highly rated hydraulic lines and heavy-duty pumps. Do not purchase a plow that is too light-duty for the intended use. Saving a few dollars at the start can be very expensive in the long run.

Furthermore, repairing a plow during a wind-driven snowstorm, with temperatures hovering around zero, can be very uncomfortable. Be cognizant of such things as the size of the quick disconnects on the hydraulic hoses, the strength of these hoses, the size and location of the connecting pins and the ease of operation from the truck cab. Even with these parts' increased reliability, accidents still happen. So be prepared.

Plows now come with various quick attach and detach systems to allow for easy connection and disconnection from the truck. Here, too, technology advances have added reliability and dependability to these conveniences. When I wrote the first edition of this book in 2002, the industry was undergoing an innovation revolution. Many of these new items, inventions and ideas were in their infancy and needed further work to achieve a consistent level of reliability.

To Salt Or Not To Salt? Another service most snow professionals add as

they grow is deicing, whether it is a simple salt spreader mounted on the back of a pickup truck or a sophisticated liquid deicing application system. Most entry-level contractors mount a salt spreader to the back of a pickup truck and fill it with bagged rock salt or other ice-melting products. For contractors doing commercial snow plowing, deicing is a "must-have" service offering. Commercial and retail customers often demand their sites be kept safe, and deicing is the only real way to make that happen. The margins on deicing are very favorable, and it is easy to justify the expense associated with securing this equipment.

Employee Management. Effectively overseeing all aspects of the business becomes virtually impossible as the business grows in size and revenue. No matter how many hours you put in, you simply cannot do it all. This means that hiring and training employees becomes an integral part of your success and/or failure.

Finding and keeping qualified employees is not easy if you offer low wages, skimpy benefits and minimal room for advancement. When starting out, you may not be able to offer such incentives, but you will need to strike a balance between what you can afford and what you cannot fail to provide. This is one important reason why employee relations are a key part of small business management. Employees of small companies are often more satisfied than those working at larger firms. Often their ideas are better received and they feel they are making a difference. The small business owner often respects his or her employees more, knowing "they are all in this together."

However, responsibility and delegation of authority can be a touchy issue when dealing with small-business owners. Long-term employees can be difficult to deal with as the company grows and the time comes to consider promotions. The small business owner feels loyalty to those who have been with him the longest, but unfortunately that is not a justifiable reason to heap authority and responsibility on that person. While you might feel compelled to do this, you must, for the betterment of the company, step back and ask: "How will this help the company overall?"

The same can be said for family-owned-and-operated businesses. The old saying that "you can't fire family" is just that – an old saying. Owners of small businesses can best serve themselves and their businesses if they gradually recruit and groom employees for management positions from outside their family trees. Eventually, you may have to allow your business to graduate from an entrepreneurial company into a managerial one.

Documentation. Documentation is the one thing that differentiates the small, struggling business from the successful, thriving professional. Too often, an entrepreneur says he or she is too busy to keep track of everything. In the snow industry, record keeping is paramount to success.

Successful snow management companies are very thorough about documentation. They track where people are working, when salt trucks arrive on site, how much material they use, phone calls that come into the business requesting services and even outgoing phone calls to customers. Other information that needs to be documented include snowfall depths at various customer locations, temperature fluctuations during a snow or ice event, how many cars are in the customer's lot, special requests made by customers and key customer contact information.

As your company grows and becomes more involved with insurance companies, adjusters and customers who question invoices, you will realize the importance of recordkeeping. Detailed, accurate records can mean the difference between success and failure. Software advances allow for more accurate recordkeeping and assist snow contractors in running their business more efficiently. In fact, progressive contractors can now track every piece of on-site equipment, the amount of work it completes and the cost associated with the labor and/or equipment. This can now be put directly into an accounting system that generates customer and subcontractor invoices for each push or event.

Working As A Subcontractor

Once you have the necessary equipment, there are several ways to enter the snow and ice management industry. One common entry method is to become a subcontractor for an established company in your area.

Working as a subcontractor is relatively easy and hassle free. You generate income for yourself and you don't have the headaches associated with managing a business. You don't have to sell yourself to customers, price jobs, invoice customers, or worry about turning a profit from sales and services. As a subcontractor, you do the work and get paid a fair amount for your efforts.

Working as a subcontractor also provides several benefits that come with

SNOW FACT:
The greatest snowfall officially reported at the Phoenix, Arizona, National Weather Service Office was 1 inch. That occurred twice. The first time was January 20, 1933. It happened again four years later on the same date.

self-employment. You can depreciate your equipment and deduct associated expenses. These legal deductions allow you to keep a lot of the money you earn. However, always consult a tax professional on what is permitted.

Some subcontractors do quite well working for other snow contractors. A loyal subcontractor who enjoys plowing or clearing ice can generate a nice income working under the direction of a larger contractor who takes all the risks, finds all the work and handles all the documentation.

Finding The Right Partner

Finding reputable contractors to work for is not as easy as it sounds, and has become even more problematic in recent years. Unfortunately, there are a few unscrupulous contractors who promise the world and then fail to deliver. This has occurred more frequently with the emergence of so-called "national providers." However, most locally or regionally based contractors still trade on their good name and reputation.

When you seek out a company to subcontract for, research it as you would any potential employer (although you will not be an employee when you work as a subcontractor). The best reference sources when seeking a company to plow for are the subs who are already working for the contractor.

Usually, you will hear by word-of-mouth or through an advertisement in the "General Help" section of the local newspaper who is hiring. You can also research a company by visiting and studying their website. However, there is no substitute for doing some old-fashioned detective work and contacting other subcontractors.

"Interviewing" a potential contrac-

tor to work for is also a good idea, but remember, the interview process is a two-way street. The contractor will ask you questions and check your references. When preparing for the meeting, consider these questions:

- How do you route your subcontractors?
- What kind of customers does your company service (residential, commercial, retail, industrial)?
- How do you pay your subcontractors?
- What kind of insurance do you require of your subcontractors?
- Do you plow gravel lots?
- How much can I expect to work during any given snow event?
- Do you run a "speed crew" for minor events?
- How will I fit into the "speed crews" and/or clean-up crews you use after a storm is complete and most lots are cleared?
- What about breakdown time?
- What is your policy on travel time between jobs?
- Can I plow some of my own accounts, too (not while you are paying me, of course, but I do want to grow my business, as well)?
- What is your policy about who gets called out first?
- Will you let me talk to some of your other subcontractors?

A viable and honest contractor should have no problem providing references for this outfit. Call a few of the references and chat about the contractor's policies. Investigate what technology he employs for tracking your movements, as this can tell you how progressive the company is as they con-

INNOVATIVE SNOW CONTRACTORS BRING IDEAS TO THE TABLE

As with most things in the snow management industry, many of the recent industry-related advancements have come from snow contractors themselves. Entrepreneurial snow contractors focused on coming up with advancements to solve some of the key issues have given us most of the advancements in equipment and technology. One such "invention" is software to track what subcontractors and employees are doing in the field during a snow or ice storm.

Daniel Gilliland of Snowfighters.US from Belton, Mo., is one such snow professional. Having been involved in the snow industry for many years, he knew there had to be a better way to track production practices. Gilliland pioneered a software system (www.crewtrackersoftware.com) that allows employees and subcontractors working sites to call in and record the work that has been completed. This software interacts closely with QuickBooks allowing the progressive snow contractor to know (in real time) what services have been completed, the estimated direct costs and the estimated revenue that will be generated.

By interacting with QuickBooks, subcontractor invoices can be generated and recorded the day the work was done. Customers can be invoiced that same day and the associated accounting functions can be completed with margin reports generated automatically. Add to this the ability for customers to see (again, in real time) what has been completed and you have a useful tool that was previously unavailable to the industry at large. Of course, there are others industry contractors who have successfully implemented innovative technologies allowing them to speed growth while still having adequate back-end systems in place to track and document vital information.

tinue their own growth strategies.

A good sign is when you can find subcontractors who have worked for the contractor for several years. Make sure the contractor has a good reputation among his peers, and make sure there are no unresolved issues existing with the local office of the attorney general or Better Business Bureau.

Snow As An Add-On Service

In the past, it was the rare individual who started his or her career as a business owner in the snow management industry. More often than not, an in-dividual contemplating entering the snow management industry already owned a service-industry business. The most common of these included landscape, pavement maintenance, excavation, roofing and pool installation contractors, tree care companies and other seasonal-type contractors.

Sometimes these contractors entered snow management because they need ed to keep existing customers happy, keep valued employees active during winter months and generate revenue.

Today, snow and ice management is viewed as a profit center, and not as a necessary evil. If your core business has some of the necessary equipment and staffing available, it may make sense to add snow management to the mix. However, some issues should be resolved within your business prior to making this decision.

Staffing. The staff of your existing business is one of the first consider-

> ## "There are **contractors who are quite successful** in operating businesses that do snow and ice management exclusively."

ations. Will employees support such a decision? One scenario leading to a successful introduction is to allow an upper-level management employee to research the possibilities before implementing a snow business model.

If the support staff buys into the idea before you issue a mandate, the entire process of adding snow management will be easier. The arguments against such an endeavor usually stem from lack of knowledge about what is actually involved with entering the snow management business. To alleviate this problem, properly inform your staff about the pros and cons.

The Power of Networking. Networking with other successful snow contractors, through associations and industry trade shows and events, can give you insight about venturing into the snow management industry. Such involvement also generates excitement for the process and alleviates the fears associated with any new endeavor. As an add-on operation, a snow removal and ice control operation will generate higher gross margins than your core business, but only if approached properly.

While still the exception rather than the rule, there are contractors who are quite successful in operating businesses that do snow and ice management exclusively. It used to be these contractors were typically located in geographic areas with heavy snowfalls, and they almost always had a large pool of commercial clients.

However, some of the most profitable and successful snow-only contractors are in areas previously not considered snow markets, such as Maryland, New Jersey, Kansas and Virginia. It is possible for contractors to make a healthy profit, though only generating revenue for three to five months.

Technology and entrepreneurialism have lowered the entry barrier into the snow-only arena. And while the combination of enhanced business education and advanced technology have made this possible, competition has also encouraged growth.

For such business models to work, contractors must have deep relationships with reliable subcontractors who own equipment designed to service large-scale commercial/industrial clients. Otherwise, the contractors have to own the equipment, which makes earning a profit nearly impossible, especially since the equipment will sit idle during non-winter months.

These factors – reliance on subcontractors, need for long and heavy winters and a roster of deep-pocketed commercial clients – make owning a stand-alone snow business a serious undertaking.

NOTES:

Chapter Three
BECOMING A BUSINESS

Chapter Highlights

- Types Of Businesses
- Partnerships vs. Sole Proprietorship
- Corporations
- Limited Liability Companies

When you want to establish yourself as a business and operate independently, one of the first things on the to-do list is obtaining a federal employer identification number (EIN). An EIN is a nine-digit number the Internal Revenue Service (IRS) assigns to employers for tax filing and reporting purposes. You do not need to have employees to obtain an EIN. EINs are free and they are easy to obtain. You may use your EIN on all tax returns, employment tax checks, employment tax returns and other employment documents that you send to the IRS.

You can obtain an EIN by filing IRS Form SS-4 (Application for Employer Identification Number) with the IRS (visit www.irs.gov). Completing the form is simple, and detailed instructions are included. You can obtain your EIN by mailing the completed form to the appropriate IRS service center listed in the form's instructions. The IRS will mail the EIN to you in about four weeks. If you need the EIN right away, you can get it via telephone by using the IRS's Tele-TIN program. The SS-4 form has instructions for this.

Having an EIN establishes you as

a business, and some companies you work with will request this information. Your EIN number can be used on the standard W-9 form you may have to fill out when companies pay you for services rendered. Recent changes to the law now require all companies to collect the standard W-9 form if you are paid over $600 in any given reporting year. It no longer matters if you are incorporated.

Previously, if you were a corporation, you did not receive a 1099 (which is generated by whomever pays you money, and they need the W-9 on file to generate a 1099). You use the EIN even if you are a sole proprietor. The EIN would be used in place of your Social Security number on the W-9. Some companies will look more favorably on your status as a business if you have an EIN. (*Author's Note: At the time the 2nd edition of Managing Snow & Ice was being written, a significant movement was afoot to have the law that corporations supply 1099s, repealed.*)

Being self-employed has many advantages, but there is a significant time commitment involved in doing so. Customers can be quite demanding, and the work we do in the snow industry is often done at difficult hours, under horrendous conditions, which takes time away from friends and family.

Anyone who tells you that being self-employed is fun, you can take time off whenever you want, and you'll have less worries has never experienced a growing business that is "your baby." If you get sick, get hurt or become incapacitated in any fashion, the business suffers from your absence.

Many a snow and ice contractor has lamented the lost time with family when it snows on Thanksgiving, Christmas or New Year's when a snow or ice storm moves through town. It comes with the territory in this business, and it is the rare individual who can actually spend significant time away from his or her business and *not* worry.

After several years of growth, combined with the hiring of competent staff, confidence can be gained allowing others in the company to make significant decisions with regard to day-to-day activities. However, this takes time and diligence, and most self-employed individuals choose to stay close to their businesses at all times, especially in the early years.

Types of Businesses

There are three basic business entities available to the snow and ice management professional when starting up his or her business. They include:

- Sole Proprietorship
- Partnership
- Corporation

Each entity has its advantages and disadvantages, and we will discuss each in greater detail.

Sole Proprietorship. A sole proprietorship is an organization that is owned and operated by one person. It is the most common form of business ownership in the United States. A sole proprietorship is easy to start and end. All you have to do to start a sole proprietorship is simply say, "I'm in business." The simplest form of announcement is satisfactory evidence that you are in business. The business is just as easily dissolved. You simply stop. There is no one to consult or to disagree with about such decisions – although your spouse and family usually have some input. There is pride in ownership and you do not have to share the profits with anyone else – except for mandatory taxes to Uncle Sam.

Interestingly, while sole proprietorships are the most common form of business ownership in the U.S., they only represent 5% to 10% of the sales volume. There are no special taxes associated with being a sole proprietor, and profits are taxed as personal income. However, the owner does have to pay estimated taxes quarterly, and you cannot wait until April 15 each year to settle up with the government.

The down side is you cannot collect unemployment if times are slow and

There are three basic types of businesses for snow contractors to consider - sole proprietorship, partnership, and corporation.

cannot deduct your health insurance premiums as business expenses. There is no sick leave or paid disability insurance, and these fringe benefits for paid workers/employees adds as much as 30% to their total wage compensation.

There are other disadvantages to being a sole proprietor. For example, when you work for another person, it's his or her risk that is problematic. As a sole proprietor, *you* assume all the risks of being in business and debts you incur are *your* responsibility. This means if you mismanage the business, you can lose your home, your savings or your personal assets (cars, boats, etc.). This is a serious consequence of being self-employed and not one that should be taken lightly.

Should you die unexpectedly, the business ends unless your heirs take over or unless you have made provisions for the sale of the business after your death. Often, though, your customers are dealing with you, the sole ownership entity, and will not want to deal with anyone

else should you die. This is a concern if you, your family and employees have begun to achieve wealth through operation of the business.

This is also a very good argument for allowing others to have leadership roles in your company. Often, "success" is defined as allowing others to also grow personally and professionally in the business. It is one form of safeguarding the business after your demise. Having someone in place who can carry the torch and serve as the face of the company allows for continued operation after your untimely departure.

Partnership. A partnership arrangement can have advantages over the sole proprietorship, but it too has its drawbacks. There are several types of partnership arrangements to consider including general, limited and master limited partnerships.

General Partnership. The general partnership is one in which all owners

33

share in operating the business and in assuming liability for the business.

Limited Partnership. A limited partnership is a partnership with one or more general partners and one or more limited partners. The general partner is an owner who has unlimited liability and is active in managing the firm. A limited partner risks an investment in the firm but enjoys limited liability and cannot

> "Do not jump into a **partnership relationship** without considerable due diligence."

legally help manage the company. By "limited liability" we mean that the limited partners are not responsible for the company's debts beyond the amount of their investment. Their liability is limited to the amount they put into the company, and their personal assets are not at risk.

Master Limited Partnership. The master limited partnership acts much like a corporation and may be traded on the stock market. However, a master limited partnership pays taxes like a partnership. In this way the business avoids the required corporate income tax. It is rare to see this type of entity in the snow and ice management industry.

Partnerships vs. Sole Proprietorship

There are a few distinct advantages of the partnership arrangement over the sole proprietorship entity. For one, there is a pooling of resources that comes with having a partner. Some would contend it is easier to manage a business with a partner, a person who you can share information and management responsibilities with.

There are shared financial resources to be considered, as well. The liability is shared, thus taking the burden off one individual to provide the financial wherewithal to finance growth, acquisition of equipment and inventory, and make other key decisions. Management of day-to-day activities is shared. Therefore, the overall workload is distributed amongst the partners.

There are disadvantages to partnerships, as well. Anytime two or more people must agree on something, tension can ensue. Partnerships can destroy families, friendships and marriages. You must share the profits with others, even if one of the partners believes he is doing most of the work. Liability is shared, which means if a partner makes a costly mistake through ignorance or an error in judgment, all the partners will take the hit. General partners can lose their personal assets and not be the one at fault.

Disagreements among the partners can be the cause for dissolution of the company. If such dissolution takes place, there can be considerable debate over who gets what. It is best to decide – in writing – prior to going into business together how the entity will be split up in the event of dissolution. It's almost like having a prenuptial agreement, which in fact, a partnership is very much like. Do not jump into a partnership relationship without considerable due diligence. It is time well spent, especially if you have to end the relationship.

Corporations

While the word "corporation" makes

one immediately think of Microsoft or IBM, small businesses often opt for corporate status as a business entity. A "C" corporation is a state-chartered legal entity with authority to act and have liability separate from its owners. This means stockholders (owners) of the corporation are not responsible (liable) for the debts or any problems of the corporation beyond the money they invest into the corporation. As a result, the stockholders no longer have to worry about losing their homes and personal assets.

A corporation not only limits the owners' (stockholders) liability, it also allows many people to share in the ownership (and profits) of a business without working there or having any other management commitments to the business.

In most snow and ice management companies, the stockholders are the owners and often do the majority of the work necessary to keep the business operating. This limited liability is the major reason for establishing a corporate status.

Additionally, the corporation becomes a perpetual entity that can survive the death of any of the stockholders. It is also easy to change ownership of the entity as a stockholder simply sells his or her stock to an interested buyer. And, it is possible to separate ownership from the management of the corporation.

There are disadvantages to consider when choosing a corporate setup. The initial cost can be expensive if you involve lawyers and accountants. However, in some states there are now easier and less expensive ways to do this yourself.

The recordkeeping requirements in sole proprietorships are less stringent. For a corporation, there are tougher rules to follow when it comes to documentation. These rules include filing two tax returns – one for the individual who owns the stock in the corporation and another for the corporation. The corporate tax return, which can be quite complex if the corporate entity is large, can result in double taxation for the individual and select states tax corporations at a higher rate than other entities discussed here.

Keep in mind a corporation does not require numerous owners or hundreds of stockholders. Individuals may incorporate themselves to take advantage of the liability protection that is afforded those who hold stock in such entities. Many firms incorporate in Delaware simply because the application process is easier to complete versus other states.

An excellent resource to learn about the steps needed to establish your business as a corporation is *"How to Form Your Own Corporation without a Lawyer for Under $500,"* by Ted S. Nicholas

A corporation protects the owners from certain financial liability but requires additional recordkeeping.

When setting up your business, you might want to consult with your accountant and lawyer to decide which format is best for you.

(published by Enterprise Publications).

"S" Corporation. An "S" corporation is a unique government creation that looks like a corporation but is taxed like a sole proprietorship. "S" corporations have the benefit of limited liability, and the "paperwork" is much like a regular corporation. They have shareholders, directors, employees, etc., but the profits are taxed as the personal income of the shareholder(s). This avoids the double taxation of regular corporations.

"S" corporations must pay self-employment taxes in the same manner as sole proprietorships, but not every business can become an "S" corporation. To qualify as an "S" corporation, a company must:

- Have shareholders who are individuals or estates, and are citizens or permanent residents of the United States.
- Have only one class of outstanding stock.
- Not own 80% or more of the stock of another corporation.
- Not have more than 25% of income derived from passive sources (rents, royalties, interest, etc.).

Limited Liability Company

A limited liability company (LLC), also known as a company with limited liability (WLL), is a flexible form of enterprise that blends elements of partnership and corporate structures. It is a legal form of a company that provides limited liability to its owners in the vast majority of United States jurisdictions. LLCs do not need to be organized for profit. Often incorrectly called a "limited liability corporation," it is a hybrid business entity having certain characteristics of both a corporation and a partnership or sole proprietorship (depending on how many owners there are).

An LLC, although a business entity, is a type of unincorporated association and is not a corporation. The primary characteristic an LLC shares with a corporation is limited liability, and the primary characteristic it shares with a partnership is the availability of pass-through-income taxation. It is often more flexible than a corporation, and it is well-suited for companies with a single owner.

Limited liability companies are becoming very popular for contractors in the snow and ice management industry. It is important to understand that limited liability does not imply that owners are always fully protected from personal liabilities. Courts can, and sometimes will, pierce the corporate veil of corporations or LLCs when some type of fraud or misrepresentation is involved.

As you can see, there are a number of options for your business. Work with your accountant and attorney to choose the one that best fits your situation and business goals.

NOTES:

Chapter Four
FINDING CUSTOMERS

Chapter Highlights

- Advice For Startups
- The Customer Meeting
- Customer Surveys
- Marketing For Established Companies
- Relationship Marketing
- Promotion And Advertising

No business exists without customers. So for any snow and ice management professional, new or experienced, finding customers should be a major focus. While recent advances in technology have afforded contractors an easier way to communicate with potential customers, some of the old, tried-and-true methods still work. When the two are combined with the advances technology has brought us, some amazing things can happen with regard to "getting the message" out to customers.

Advice For Startups

Once the decision has been made to enter the snow and ice management industry, and after acquiring the necessary equipment, one needs to secure customers to generate revenue. You have to find customers and get them to notice you. Straightforward approaches usually work best.

Canvassing Neighborhoods. One method is to target a neighborhood and distribute promotional pieces that highlight your company's winter services. Placing several hundred pieces in a neighborhood that appears to be af-

An effective sales brochure can help promote your business to new customers.

fluent enough to warrant snow and ice management services is an ideal place to start. Include a short description of the services offered, your name or the name of the business entity, a phone number, email address and a website address (if you have one). Remember, you cannot put these pieces in mailboxes. This is strictly forbidden by the U.S. Postal Service, and can get you into some considerable trouble with federal authorities.

Timing is important with residential customers. Potential customers are most likely not thinking about snow-plowing services in July when they are swimming in their backyard pool. Waiting until the temperature drops can increase the interest in your services, as the declining temperature is likely to bring winter thoughts into your potential customers' minds. Be ready to "get the word out" the day after the first really cold night in late October or early November. With residential customers, direct contact or referrals from friends and neighbors usually works best.

Of course, if your goal is to secure a contract with a homeowners association (HOA), condominium or apartment association, the strategy you would use for contacting them would be different. These associations normally use a management company to oversee the bidding and selection process for maintenance and upkeep services with outside contractors. These management companies have websites and often are affiliated with local branches of national organizations.

Using available search engines such as Google or Bing, snow and ice contractors can perform simple searches for their particular market area and obtain leads. Reaching out to local management companies for work is a good way to make contact with people who are looking for service companies to choose from and work with. It can be as simple as asking to be put on their list of qualified bidders.

Of course, as you start to develop contacts and promote your company in the market, you may have to demonstrate that you are a "real business," not a fly-by-night operator. Aside from the various legal and operational startup strategies outlined in Chapter 3 *(page 31)*, one simple way to raise

your company's professional stature is by creating a website.

When getting started, a self-made website (there are various off-the-shelf design programs you can purchase) may be adequate. However, projecting a professional image is a must in today's world. Consider hiring a graphic artist with website design experience to create your company's site.

To that end, as the company owner, accept and understand that having a professional-looking website can be just as important as dressing properly for a face-to-face appointment. A website is a great way to impart information about what services you offer, what geographic area you service, your equipment stable and how you deliver snow and ice management services in a timely, professional manner. Anyone can make up a slick website. But in today's world, a website is viewed as standard business practice, not a luxury. Operating without a website, email and social media (Facebook, Twitter, LinkedIn, etc.) can be viewed as unprofessional, unsophisticated, or just plain backward.

The Customer Meeting

Once calls or emails start coming as a result of your marketing efforts, make appointments with potential customers to educate them about your company and the services you offer. Create business cards or collateral promotional materials detailing your company's facts, features and service offerings to leave behind with customers. Business cards are inexpensive, and you can print them yourself with software that is readily available at local office supply stores.

Always be honest with potential customers. When speaking to a prospective customer, explain that you are just starting out. If there is any resistance, accentuate the positives of working with a small company – personal service, ability to respond quickly to requests, the owner is hands-on, etc. You should also practice the "pitch" in the shower or in front of a mirror. It sounds hokey, but all great salespeople "practice" – in the car, in front of others or just running the scenario through their minds when alone. Any practice is good preparation for when you are sitting in front of the customer.

Beware of the potential customer who tells you, "If you give me a good price, I'll get you a lot of work in this neighborhood." Rarely do these promises come to fruition. Fair and honest practices will get you more new business than low prices. Always price your jobs to make a fair and honest profit. Some new snow contractors get hung up on what others are charging and then attempt to "match the price."

Only one company in the world has thrived by offering and giving customers "the lowest price – guaranteed." However, they have the ability to purchase high-volumes of goods at such low rates, offering low prices has become a staple for them. It is the rare snow and ice contractor who gets – and keeps – business year after year by being the cheapest in the marketplace.

Marketing For Established Companies

Finding customers for your snow and ice management services is less difficult if you have an existing customer base to work with. For example, a customer database from a landscape installation, maintenance or lawn care company makes the ideal entryway for marketing your snow and ice management services.

Think you can save money by letting nature take its course and do nothing about clearing that snow off your parking ramp?

Think again.
Think death. Think damage. Think liability.

Gravity is a law of nature. Don't break it!
Call SNOWSCAPING PROPERTY MAINTENANCE now and get that snow off your ramp before penny-pinching leaves you penny-less!

SNOWSCAPING PROPERTY MAINTENANCE **800-555-4321**

Creating an effective, detailed proposal helps your sales efforts.

The first step is to contact all of your existing customers. Studies have shown that a high percentage of any good service business' new revenues come from existing and past customers. It is also common knowledge that it takes fewer marketing dollars to keep existing customers and generate new business from them than it does cultivate find new customers.

A simple promotional piece listing your company's snow and ice management services is easy to generate and distribute to customers. Use a mailing to generate interest with past customers – assuming you keep complete, accurate client records. It should include a personal letter from you, the promotional piece and possibly a discount coupon for being a loyal customer if they sign up in the next 30 days.

Name recognition marketing is the best form of brand awareness. One of the best ways to secure attention from potential and existing customers is to use electronic postcards (e-cards). An e-card can be developed for under

41

$300 and there are no printing costs associated with this type of marketing. You can make contact with thousands of potential customers with just a few hours in front of a computer.

Of course, sending out a single e-card may accomplish nothing. Instead, a targeted email campaign to gain brand awareness is best. Marketing studies indicate you need as many as five "touches" with a prospective customer before they recognize your name. Staggered, repeated sends of e-cards can accomplish this, but be wary of "attaching" postcards to an email. The postcard needs to be embedded in the body of the email itself. That way, once the email is opened, it becomes readable with little additional effort.

Exercise caution, though, when pursuing digital marketing through email. To avoid being branded a "spammer" you must allow the recipient the opportunity to opt out of receiving any additional email advertising. An "unsubscribe" link must be included (usually at the bottom of the e-card) that allows the recipient to tell you to stop sending emails. When you receive such a request, you must remove them from your list immediately. While there are various contact management software programs available to assist with this kind of marketing, you can do it yourself with Microsoft Outlook and an Excel spreadsheet.

If you have a sales staff, they should follow up by phone with current or former customers to introduce the services. If they are already under contract for snow removal, ask more questions. Have a form ready for such calls so you can gather information that will be helpful to you in future sales calls.

What types of questions should you ask? Start with, "Are you satisfied with your current service provider?" If they are, press further and tell them it is good they are satisfied with their current company, but ask if they would help you learn how best to provide good service to other customers.

For example, say, "That's great. I'm glad you are satisfied with ABC Snow & Ice Removal and I won't try to pressure you into switching contractors. However, we want to provide the same high quality service you are currently accustomed to. Can you help me by educating me on what you consider is a good snow contractor?"

Assuming an affirmative response, continue to ask the following questions:

- What level of service do you demand?
- How much are you paying for snow and ice management?
- What kind of plowing equipment does your current provider use?
- Who is this good contractor?
- What are their billing procedures?
- What makes them a good contractor?

E-mail marketing is an effective way to promote your snow and ice removal services.

You can rely on Winter-Pro Management Snow Removal Services to be your hero during winter storm events...

...but don't just take our word for it:

Winter-Pro Management has preformed anti-icing and snow removal services fo[r] [P]arking Lots Ltd. for the past 7 years. Bob and his team have provided exceptio[nal] [s]ervices for my portfolio. The quality of work and professionalism is unparalleled, [h]ence the continued working relationship that we've developed. I would certainl[y] [re]commend Winter-Pro Management for snow removal services without hesitati[on].

[L]arry Slevinson, Owner
[P]arking Lots Ltd.

WINTER-PRO MANAGEMENT
"Where You Go To Remove Snow"

Call 860-555-6798 Today!

Mark down and save the answers as they are important for several reasons. You are developing a picture of what this customer perceives a good snow and ice management contractor to be, and you are finding out pricing structures. In essence, you are doing a marketing survey.

Customer Surveys

Another avenue to gain information is to send a questionnaire to existing customers asking them if snow and ice management, as an add-on service, would be of value to them. If this is presented as an information-gathering endeavor, and not a solicitation for business, you may be pleasantly surprised by the response. People love to be asked their opinions. They often want to help and are easier to approach than most of us think.

Networking. When it comes to networking, it would be wise to seek out the best snow contractor in your market and talk. People love to talk about themselves. If you approach a reputable snow contractor, the chances are good he will be much more open to you if you tell him up front that you are considering entering the business and would like his advice.

Be honest that you won't be able to compete with him at his level. If it should happen, you would like to be a good competitor and not branded as a "low-baller." In seeking the advice of an experienced person in the industry, you automatically set yourself apart from the "plow jockey."

Most businessmen would much rather educate the person just starting out, as it will keep you from degrading the industry with poor business practices.

You should also consider asking larger contractors in your market for referrals. All large snow removal contractors stop taking on new business at some point, as it is good business not to over book. It's also good business to refer inquiries to another viable contractor if you cannot take the account.

If you approach larger contractors with this in mind, you may end up with a tremendous book of business in a very short time. Some larger contractors can make small contractors viable just through referrals. If this happens, don't forget who helped you when you were small. Favors should be repaid at some point in the future.

Often, larger contractors seek out subcontractors who are reliable, professional and will show up and do a good job. Consider working as a subcontractor for a winter or two to learn the industry and grow your customer base. You are well within your rights to "check out" the contractor who hires you as a subcontractor *(See Chapter 3, Becoming A Business, page 31)* and why. You want to make sure you get paid in a timely fashion for the work your company performed.

There are other practical avenues to secure new customers and referrals including handing out business cards to people you meet at church, your kid's ballgame or school event, or at the neighbor's cookout. Asking for referrals takes little effort and is often overlooked during the service process. Most people won't ask for a referral simply because they have thin skin. Remember, the word "no" will not hurt you.

And don't forget that old standby – the cold call. Soliciting business is not bad business. Just keep in mind that this approach takes time. You will get turned down and told no, but asking for business is an age-old sales tool that works.

SNOW FACT:

New York City used more than 300,000 tons of salt during the 2010-2011 winter season for deicing.

Source: New York State Department of Transportation

We will deal with some specific sales strategies for snow and ice management services in Chapter 5 *(See page 51)*.

Other Marketing Methods. You must establish your company's own identity and determine how you will service your customers. Once that is determined, communicating this message to potential customers is defined as "marketing."

Marketing is many things to many people. Most small businesses deal with marketing as a means to an end. The end result is to secure new business and retain it for the long-term. Marketing is not sales, nor is it a substitute for a good sales presentation. Marketing is the process of determining customer wants and needs and then providing the customer with goods and services that meet or exceed their expectations.

Defining the wants and needs should be fairly easy. However, providing the customer with the service that meets those expectations takes more work. The challenge is to define what any particular customer's expectations might be when it comes to snow and ice management. Some customers, the industrial sector, for example, may not require bare pavement, as they only need to get employees into the manufacturing building and don't have constant traffic.

Retail customers have the greatest liability exposure and the most traffic into their facilities. Therefore, they typically require a higher level of service. Some customers require 24-hour-a-day, 7-day-a-week service. This means a tremendous commitment on the part of the snow and ice management contractor, as equipment may need to be manned all the time in the event of a catastrophic snow or ice event. Your own personal commitment level determines what level of service you can – or will – supply to customers.

With the previously mentioned facts in mind, some decisions need to be made. What level of service do you desire to provide customers? Having your name or the company's name known for professional, reliable and quality service is much more desirable than being known for cheap prices. Nothing of value comes cheap, and having an identity that denotes cheap should not be anyone's goal in this industry. If you can lower a customer's expenditures by achieving better productivity, this might be a viable reason for potential customers to deal with your company. However, one must have an identity in order to have a reputation.

This is where marketing comes into play. Marketing your company's snow and ice management services establishes an identity. If the goal is to have a reputation and identity that you provide quality service and protect your customers' liability exposure (and their customers' in the case of retail establishments), then you need to market your company along those lines.

The most important thing is to be able to back up what you promise customers. For example, establishing a reputation for delivering bare-pavement service takes time. In many cases, especially in tough economic times, bare pavement service is sacrificed in favor of more advantageous pricing structures *(See Chapter 7, Pricing Snow Work Profitably, page 69)*. However, you can assist that effort through effective marketing of your company's service standards.

Relationship Marketing
Relationship marketing was first developed from direct-response marketing

Getting involved and supporting community projects is a wise investment for businesses.

campaigns that emphasized customer retention and satisfaction, rather than a dominant focus on gaining new sales.

As a practice, relationship marketing differs from other forms of marketing in that it recognizes the long-term value of customer relationships and extends communication beyond standard advertising and sales promotional messages.

With the growth of the Internet, relationship marketing has evolved as technology opens more collaborative and social communication channels. Just like customer relationship management (CRM), relationship marketing is a broadly-recognized, widely-implemented strategy for managing and nurturing a company's interactions with clients and sales prospects. It also involves using available technology to assist and organize one's efforts to gain and retain customers.

The overall goals are to find, attract and win new clients, nurture and retain those the company already has, entice former clients back into the fold and reduce the costs of marketing and client service.

Snow and ice management services fall under this definition because most of our potential customers have specific wants and needs. By utilizing relationship marketing practices, you can work at retaining customers instead of constantly finding new ones. Retaining customers is much more favorable than constantly turning over clients in the hope of finding new ones.

Some professionals believe that retaining an additional 2% of their customer base has the same effect on profit as cutting costs by 10%. Relationship marketing means establishing and maintaining long-term relationships with customers.

You cannot take a shotgun approach to marketing your company. You cannot be all things to all people. Decide if you wish to concentrate on the residential or the commercial markets. We shall discuss the merits of both later in this book *(See Chapter 5,*

Sales, page 51), but it is a decision that must be made at some point.

Promotion & Advertising

The established promotional tools available to a snow and ice management company are typical of most service-industry outfits. The list can include:

- Print and Internet advertising
- Word-of-mouth
- Sales promotion
- Public/community relations
- Social Media and digital marketing (websites, email, blogs, etc.)

Many people equate promotion with advertising because they are not aware of the differences between them. Advertising is paid, non-personal communication through various media by companies and individuals who are in some way identified in the advertising message.

Word-of-mouth is not a form of advertising because it doesn't use an alternative medium – newspapers, Yellow Pages, radio, television, etc. It's not paid for and it's personal. Publicity is different from advertising in that media space for publicity is not paid for. Personal selling is a face-to-face style of communication and doesn't go through another medium, and thus, is not advertising.

Advertising. Advertising is a form of communication intended to persuade a typically larger audience – viewers, readers, or listeners – to purchase or take some action upon products, ideas, or services. The message includes the type of service and how that service could benefit the consumer.

Advertising can also serve to communicate an idea to a large number of people in an attempt to convince them to take a certain action. Advertising is a method of "getting the word out." For years, Yellow Page advertising was the way snow and ice management professionals promoted their services. Recent advances in technology have made "telephone books" a dying breed, and Internet-based lists – for example, craigslist and Angie's List – are popular. Advertising in the newspaper can narrow the focus of your efforts, whereas television advertising can widen the scope. However, television can be inefficient from a cost-analysis standpoint as it is usually expensive and may not offer an adequate return on your investment.

Internet advertising is developing rapidly and having a presence of some sort is the norm today. Internet advertising is fast replacing or greatly supplementing other advertising efforts. As was discussed earlier in the chapter, a simple, informative website helps customers get to know you, your service offerings, how you operate, the markets you operate in, and how you can best serve them. It also gives them a chance to contact you directly through email or an information request form. The website does not have to be expensive or elaborate, but it should look as if it has been professionally designed.

Having half the site "under construction" just irritates visitors. You are better off with a site that is basic, simple, and works, rather than have 100 different pages that "bury" your key information points. Images of your company in action are a powerful addition. People like to see pictures of equipment in action. Also, any website development service worth their salt will place your logo front and center on your site.

Another popular website element

GETTING SOCIAL – USING SOCIAL MEDIA TO MARKET YOUR SNOW MANAGEMENT SERVICES

Social media is everywhere today. Facebook, Twitter and LinkedIn are just a few of the outlets that millions of people are visiting on a daily basis to connect with friends and business colleagues.

This large group of consumers is very appealing to marketers of all types of products, from cars to restaurants and almost everything else in between. And since many social media users live in cold weather climates, why not market your snow and ice management services to them?

Using social media as part of your overall marketing plan makes sense and is being used more often by service industries like snow removal, lawn care and landscape contractors. But how do you go about getting started with social media and use it to your advantage? Here are some helpful tips:

Promote Your Company's Expertise – Let customers know how long your company has been in business, how many satisfied customers you have and why your plowing services are the best.

Put Your Website Content To Work – Make sure your website features customer testimonials, information about how to schedule service and obtain a quote, and a little about you and your company.

Interact With Visitors – Make sure you respond to emails and Facebook and Twitter posts in a prompt fashion. Potential customers use social media to have an interaction with someone – don't leave them hanging.

Reward Loyal Customers – You can offer discounts and promotions to customers who frequent your social media outlets and help you spread the word about your company.

Respond Quickly To Unhappy Customers – If a customer uses social media to complain about your service, respond quickly and in a positive fashion. Don't be negative in your response.

Find New Customers – Find potential customers by using keyword searches such as "snowplowing" or "snow removal" on Twitter.

Let Customers Promote Your Company – Encourage customers to post testimonials and even videos to tell others what a good job you did plowing snow from their driveway, sidewalks, or parking lots. Word-of-mouth advertising is very powerful.

that can enhance your brand marketing efforts is video. A short video featuring footage of your plows in action or employees interacting with customers can leave a powerful impression on visitors to your website. And you don't have to be a Hollywood producer to create a short 90-second or two-minute video with you doing the voice over. A high-quality video camera and video editing software can get you started.

Word-of-Mouth. Word-of-mouth is usually the most preferred method of promoting your business. This form of marketing is free, has inherent credibility and requires very little investment of time. By its very nature, word-of-mouth advertising involves people telling other people about the products and service providers they favor. When an individual recommends your company, the referral becomes an unpaid testimonial. People often trust their neighbor's advice more than a company's marketing efforts.

To achieve good word-of-mouth promotion, a company needs to focus on customer satisfaction. The company must deliver on promises made in the sales process, and focus on those customers who will, in fact, talk about the quality and value of your work. Unfortunately, most of us do not use this method of self-promotion as often – or as effectively – as we should.

Public/Community Relations. A presence in the community or your niche market through name recognition can be an important component to keeping and finding new customers. Publicity can involve something as simple as putting tasteful, yet promotional graphics on your vehicles. Promotion can also involve sponsoring a community function or a youth baseball or adult softball team. Identify your company's niche and then use those opportunities where you will be seen by your niche audience. For example, if you want to land commercial customers, consider becoming a member of and attending local chamber of commerce events.

Your company's website is an important marketing tool and often is the first point of contact with customers.

NOTES:

·

Chapter Five
SALES

Chapter Highlights

- Selling To Residential Customers
- Selling To Commercial Customers
- Sales Strategies
- Bidding vs. Negotiating
- Customer Retention

One of the major questions that arises when plowing contractors get together is, "How do I get new customers, with the cutthroat, one-truck operators out there giving away their time?" Today, contractors are also asking, "How can I compete in a world where the pricing is trending downward, without any end in sight?"

In the late 1990s and early 2000s, when the economy was chugging away nicely, the GDP was climbing steadily and times were good, this author encouraged contractors to take a stance. That stance was it might be better if we, as the contractors providing services, were in such a position that we interviewed customers who we wanted to have instead of just quoting work.

Taking the high road, instead of throwing numbers up against the wall seeing what might stick, kept the "good contractors" from playing the price slashing game. The attitude was great for the times, but as competition grew and new equipment became available that would enhance productivity, pricing began to drop. Many contractors found customers were not as enamored with "great," if "good" would lower their overall costs.

At no time has this author advocated slashing prices just to get business. However, customers, too, have become more sophisticated, and have learned more about the snow and ice management industry. As a result, they have been able to demand, and often get, lower pricing structures.

However, targeted selling is still the key to success in any business. Picking out the niche that works for you, and then focusing your efforts on it, still allows proper pricing and margins for the aggressive contractor.

Selling To Residential Customers

If your focus is to grow your business by gaining residential customers, then you have to target this market segment in a manner that will entice them to do business with you. Accentuating your commitment to perform quality work at a fair price will go a long way with residential customers that watch every dollar closely. While residential customers are generally easier to obtain, they can be harder to keep, and can often change contractors quickly. It takes a special mindset and level of commitment to properly service residential customers.

SNOW FACT:
SNOW BURST
Very intense showers of snow, often of short duration, that greatly restrict visibility and produce periods of rapid accumulation.

Preparing for and practicing your sales presentation helps you close the deal.

Residential customers want to know they are special and that you are going to cater to their specific needs. Thus, your company's marketing materials need to be geared toward that. The residential market can be quite profitable if proper scheduling is kept in the forefront of the contractor's mind. Bundling residential accounts to keep travel time to a minimum is the key to success. If you can plow a driveway in six to eight minutes (including travel between driveways) it is possible to plow up to eight customers an hour. If you wish to generate $400 per-hour, per vehicle, you can charge the customers $50 per-push and realize some terrific margins.

A successful snow contractor in Montreal, Canada, caters to more than 3,000 residential customers. This "Home Owners Association" (HOA), receives the company's undivided attention. The customers are located primarily in one geographic area, so their efforts are concentrated, with little travel time required to get from one site to the next.

In this instance, the profit margin is very good because the customers are close together. If you were to obtain seasonal contracts from this customer segment, those paid prior to the season, well, you do the math. Having the entire season's revenues "in the bank" by the time the first snow falls can make for a very favorable balance sheet.

Word-of-Mouth. Using existing residential customers as your springboard to solicit new business is an effective method of word-of-mouth advertising. Ask existing customers for referrals to their neighbors or have them provide you with testimonials. They can be written or even recorded with a mobile phone. Post these on your company's website or Facebook page – ask for permission in either case. Testimonials from existing customers work extremely well when soliciting new business. Pricing structures that are fair and honest, as well as competitive, also provide additional credibility.

An effective direct mail campaign to specific neighborhoods can also generate inquiries that can lead to new sales. A successful direct mail campaign is not just one mailing to each residence. It can take as many as three contacts to have your name recognized by the customer, so you must be diligent in your endeavor. For example, send a postcard that asks the potential customer to contact your office or visit your website to secure pricing and additional information on your snow and ice removal services.

Targeting a homeowner or condominium association, or the management companies who handle exterior maintenance for these locations, is a good way to grow your residential customer base. Today, most HOA's outsource the selection and contracting for snow and ice management services to a management company. Approaching management companies who service associations can provide significant growth with few actual "customers." Marketing to these management companies can often be as simple as calling them and asking what needs to be done to be placed on the "qualified bidders list."

Selling To
Commercial Customers

For commercial customers, a similar approach should be used. If you have a shopping plaza in a specific neighborhood, then it would be in your best interest to have additional business close to that location. It is easier for a contractor to incorporate a new customer into the schedule if they are close to an existing site you currently service.

Assuming you are providing tremendous service to your own customer, pay attention to what is happening across the street. Achieving geographic customer density is a smart play to obtain customers who are situated in close proximity.

If that lot across the street is not plowed on time, make a mental note of it. In fact, in today's technologically advanced world it is very easy to take a picture of the site with your cell phone. In the spring, send a letter to the customer and request an opportunity to speak with them about their snow and ice management needs.

You may not get the opportunity to secure that business immediately, but by staying in touch with the potential customer you should be able to arrange a meeting when the time comes to renew plowing contracts in late summer or fall.

Also ask existing commercial customers for letters of recommendation that can be used when selling to new prospects in the same area or commercial development.

You can also use the Internet to identify potential customers and prepare proposals in your service area. Online resources – such as www.goilawn.com and www.go-isnow.com – allow you to research and measure potential sites without leaving your office. You can obtain quality pictures of the site, often from all four sides, as well as overhead views. They have tools to allow you to measure pavement area, sidewalk area and even the "fall" or pitch of the parking lot to determine drainage patterns for melting snow. These tools will allow you to approach a potential customer with the information necessary to speak intelligently about their site and

SNOW FACT:

Snow kills hundreds of people each year in the United States, from traffic accidents, overexertion, exposure and avalanches.

what you can do.

If you are charging on a per-push basis, having new customers in the immediate vicinity will cut down on travel time between jobs. This is true for per-event and seasonal pricing strategies as well. If you are charging per-hour, per-truck with a minimum travel time between customers, it will mean increased revenue per truck.

Avoid taking on customers who have gravel parking lots unless you can charge for the increased time it takes to service them. You might also consider including a disclaimer in the contract that allows you to charge the customer to redistribute the gravel in the lots at the beginning of the spring.

Another method of dealing with gravel lots is to use polyurethane cutting edges on the plowing equipment. Polyurethane edges virtually eliminate the disturbance of gravel in the lot, provided the lot has had time to freeze prior to plowing operations. However, caution should be exercised when considering such accounts.

Sales Strategies

When a potential customer calls for plowing service pricing, most contractors want to know why that customer is changing vendors. It is a fair question. If the customer is unsatisfied with the service provided by their current contractor, then this is a customer worth spending time with to ascertain the reasons for being unsatisfied.

You might find that the other contractor was undercharging for his services, thus necessitating the need to short-cut the job to make a viable profit. In these cases you may need to be direct with customers and tell them they were not paying enough for the service. You are going to be more ex-

pensive, but the quality of work will be better. A potential customer who wants you to provide better service at the same price is not looking for quality and dependability – he is looking for the cheap price.

If he is simply looking to check pricing, then you may want to avoid getting involved with such tactics, unless you need the practice quoting work without getting anything in return. Price shoppers will change vendors next year or, worse yet, mid-season just to obtain a cheaper price. In this industry, the low-price contractor is usually revealed to the customer by the second significant snowfall when the level of service provided does not measure up. The caveat to this is when the contractor who has lowered the pricing structure does so by using more productive equipment, or talks the customer into lowering service level expectations in exchange for a lower price.

When talking with potential customers, don't be bashful to share your unique selling points, the quality that makes your company different and better than the competition. If you are a large contractor with a large fleet of equipment, point out there is no excuse for not showing up on time to perform the service.

Other key selling points might include having a mechanic on staff or readily available to fix broken equipment so it is back up and running in short order. You may also have a full-time customer response team that addresses special requests in a timely fashion. Share these points with the customer during the sales process.

For small contractors, sell the fact that you don't have a large number of customers and that every customer gets "personalized service." You don't have

Establishing a good relationship with customers can help promote your "word of mouth" marketing efforts.

to keep track of a large contingent of trucks, so you always know where everybody is working, and can make adjustments to schedules quickly. Personalized service means you care about that customer's needs, almost exclusively.

Remember, you are a professional. Don't hide the fact that you are making a profit in delivering snow and ice management services. While it may seem like everyone is plowing snow, the fact is you are in the minority. You provide a necessary service that requires specialized equipment, special talents and tremendous dedication to your customers. They should be glad you are there, ready and willing to provide services under terrible, and often unsafe, conditions. Sell these benefits of dealing with your company, no matter what size fleet you run.

Most importantly, keep in mind you are in the snow and ice management business year round. While most of our customers only think of snow when it starts falling for the first time that season, contractors should be thinking about snow and ice all year long.

When you are asking about the landscape maintenance business, ask about the snow management business as well. When you do a landscape, irrigation, or paver installation, or tree care service, ask, "Who does your snow management?" Is the customer happy with the service they are receiving? If they say, "Yes," then good for them. Tell them they are lucky to have a good contractor. But if they become unsatisfied, let them know you would appreciate the opportunity to present a proposal.

Also, if they are happy with their current contractor, ask what they are being charged. If for no other reason than to see what the competition is charging – you should always ask. It's no threat to the incumbent, as you know this customer is satisfied. Write down this number somewhere where it will not get lost. In a couple of years, if you get a call to quote the work, you will have some idea where the numbers are at for that particular customer.

Bidding vs. Negotiating

Large customers, who know your reputation for quality service and fair pricing, will lean toward negotiating pricing structures instead of bidding out the work. So establishing a reputation for quality work and exceptional customer service can assist you greatly in securing new business.

Approaching the pricing question with a large customer can be tricky. The customer, of course, wants to be sure the pricing is competitive. However, the final pricing structure is often a result of negotiating a fair and honest deal.

The agreement must be a "win-win-win" situation. The customer has to win, the people doing the work have to win and your company has to win. If all parties are not winners, then someone loses. Unfortunately, it seems that if one party loses, all parties lose.

Don't Be Outsold. The difference between good plowing contractors and great plowing contractors is their approach to sales and negotiation. Good contractors find ways to negotiate contracts with the customers, instead of simply bidding the package and waiting by the phone for a response.

My philosophy about sales, one that has worked well over time, is there are only two reasons to lose the sale: 1) Either you cannot do what the customer wants done; or 2) You have been outsold.

This is a simple philosophy, and if you think about it, you will find it to be very true.

Price is rarely the deciding factor in any buying decision. The price must be competitive, but there is no real value in cheap pricing. Most customers understand this, even if it is a subliminal idea that rarely comes to the surface. If price was the overriding factor in selecting a product or service, everyone would be shopping at the local dollar store.

Customer Retention

Numerous studies have shown that keeping an existing customer is much cheaper than finding a new one. With that in mind, consideration needs to be given as to what can be done to retain good customers. Even though you have provided great service at an affordable pricing structure, often a simple thank you will suffice to keep a customer happy. A thank you card may seem basic, however, a well-written note can go a long way to satisfying the customer's need to feel wanted and appreciated.

Personalized thoughts on paper show the customer you really do care about his or her business. Additionally, a personal visit during the "offseason" to say thank you will do wonders for the customer's attitude and perception of you.

In addition to personalized customer service, a successful method of retaining customers is through the use of multi-year contracts. Three-year contracts are normally sufficient in the snow and ice management industry. Customers do not have to go through the process of securing pricing each season, and you can plan more strategically for growth knowing you have an existing customer base to work from on an annual basis.

It will also save considerable time when it comes to writing renewal contracts. It means you will only have to contact one third of your customers each year to renew, leaving more time to solicit new customers and grow your company.

Can customers break a three-year commitment? Yes. However, if you are providing good service and consistent communication, customers will likely

Apartment and condominium developments have traits of both commercial and residential accounts.

stay with you from year to year.

However, there are situations where you need to temporarily adjust pricing to keep customers in your portfolio. Often, customers are required, by their own philosophy and/or by the pressures of the economic times, to push pricing downward.

Unfortunately, there are always some contractors who may consider this an opportunity to secure additional business. Such is life when competitors decide to try to level the playing field at the expense of good business practices.

When confronted with such a situation, and assuming that there is no disparity in equipment selection between you and the competition, you must make a decision. Do you match the competitor's low price or do you walk away from the project?

Sometimes, to keep a good customer, you need to temporarily lower your pricing to match the one that has been submitted to your client by the competition. As you consider this decision, do a little soul searching. Ask yourself: "Have I done all I can to prevent the competition from gaining a foothold with this customer?"

Keeping in touch with your customers becomes a sales strategy that separates you from competitors. Even in the wake of a directive from upper management to reduce expenses, loyalty allows a customer to discuss any situation that might be unfavorable to either party, before the situation rises to an unhealthy level.

Communicating throughout a snow or ice event is one way to strengthen client relationships. A simple phone call to gauge your company's performance goes a long way. It also keeps you in touch with changes that occur when your contact person leaves the company or there is a change at the on-site management firm. Staying in touch prevents you from having to play "catch-up," and having to resell yourself and your company.

Providing exceptional customer service, communicating consistently and delivering reliable, effective snow and ice management services throughout the term of the contract are the best customer retention tools for any snow contractor.

Chapter Six
INSURANCE COVERAGE

Chapter Highlights

- Defining Risk
- Risk Management
- Insurance
- Workers' Compensation
- Comprehensive General Liability
- Finding An Insurance Broker
- Bonding

Insurance coverage for snowplowing contractors is a must. Not having insurance to cover potential losses and liability is gambling with the financial futures of your employees, your family, and your company. However, understanding insurance and risk management is difficult. Most people do not understand risk management, and the ability to insure against such losses.

Managing risk is a major issue for service industry businesses today. Often, snow and ice management companies do not understand the risks they may incur. Unfortunately, a lot of the insurance companies still do not understand the snow and ice removal industry, thus making it harder to ensure adequate insurance coverage.

Additionally, snow and ice contractors cannot view themselves as simply in the business of clearing snow and ice from customers' properties. They are, in fact, risk managers themselves. Clearing snow is easy. Protecting the client, property manager, pedestrians, and vehicle drivers is much harder. Proper documentation, while not part of actual plowing operations, is just as important, if not more so, than the actual act of clearing snow and ice.

Customers rely on the snow contractor to provide a safe environment for all types of traffic across pavement and sidewalks.

Defining Risk

The term *risk* refers to the chance of loss, the degree of probability of loss, and the amount of possible loss. There are two different kinds of risk – speculative risk and pure risk.

Speculative risk involves a chance of either profit or loss, and includes the chance a business takes to make extra money by purchasing new equipment, trucks, and/or new snowplows. This also includes making business decisions in which the probability of loss may be relatively low, and the amount of the loss can be projected. Entering the snow and ice management business is an example of speculative risk because it requires capital investment and may result in a profit or loss.

Pure risk is the threat of loss with no chance for profit. Pure risk involves the threat of fire, slip and fall, accident, or loss. If these events occur, some entity incurs a monetary loss, but if the events do not occur, your company gains nothing. The risk most snow professionals are concerned with, and wish to insure against, is pure risk.

Risk Management

Once the risk is determined and identi-fied, companies have options to manage and reduce risk. These options include reduce the risk, avoid the risk, self-insure against such risk or purchase insurance against the risk.

Reducing Risk. Reducing risk includes establishing risk management preven-tion programs within your company. Such programs can include safety meet-ings, fire drills or plans on how to evacu-ate a building in the event of fire, lock out/tag out programs for equipment, mandatory seat belt use, ongoing train-ing of employees, and other accident prevention programs. Other risk man-agement tips include:

- Have a written plan and map for each account, indicating where trouble spots are, pointing out heavily traveled areas, designating where to put snow piles, etc.
- Look for expected problem areas (poor lighting, low spots, potholes, etc.) and communicate these to the property owner. Put the needed repairs in writing and take photo-graphs of the problem areas.
- Conduct background checks on employees, particularly their driv-ing records. A motor vehicle record gives you a history of accidents and violations. Don't assume that every person driving your equipment has a clean driving record, or even a valid license.
- Train and educate employees thor-oughly. Develop standard operat-ing procedures (SOPs), make sure

employees can safely operate neces-sary equipment, etc. Most accidents will come from your personnel, not you, so it's vital to make sure your staff has the same level of expertise regarding accident prevention.

- Inspect sites after work has been completed to ensure quality con-trol, and be sure you and your managers are available to clients to provide feedback, respond to ques-tions and address concerns.
- Make preventive maintenance of equipment a priority. Improperly maintained equipment can cause accidents by direct damage, or in-directly through rushed (and thus sloppy) work with a depleted fleet of equipment.
- Keep thorough and specific docu-mentation for each account. Re-cord location, start and finish times, crew members, weather con-ditions, lot conditions, etc.

Avoiding Risk. Avoiding risk is difficult – plain and simple. The most effective way, in the snow and ice management industry, to avoid risk is to not take on projects that have inherent risks as-sociated with them. Examples of ways to avoid possible risks in our industry would be to not take on projects that have a high probability of accident. Those accounts include:

- Plowing gas stations is high-risk be-cause fill caps can be slightly higher than the pavement surface.
- Parking lots with a high number of imperfections or manhole covers that stand higher than the pave-ment surface, and gravel parking lots.
- Sidewalk work because of the risk of "slip-and-fall" incidents.

Insurance

Reducing or avoiding risk is the preferable way to do business, but we all know that running a profitable business involves some level of risk that is out of your control, especially in the litigious society in which we live. Mistakes can and do happen, thus necessitating the use of insurance coverage.

An insurance policy is a written contract between the insured and an insurance company that promises to pay for all, or part of a loss. The premium is the cost of the policy coverage to the insured. Another explanation of a premium is this is the fee charged by the insurance carrier/company to cover specified losses that may occur.

Insurance companies are just like any other business – they are in it to make a profit. To assure they will make a profit, an insurance company gathers information and data to determine the extent of the risk involved.

The reason insurance companies are able to assume the risk associated with a particular policy is known as the "law of large numbers." The "law of large numbers" is the principle that states that if a large number of people or companies are exposed to the same risk, a predictable number of losses will occur during a given time period.

Another term that is important to remember is the *rule of indemnity*. This is an insurance restriction that states an insured person or company cannot collect more than the actual loss from an insurable risk. In other words, you cannot gain from risk management, you can only minimize loss. For example, a company cannot buy two insurance policies, and collect from both of them on the same loss.

There are several types of coverage insurance companies will insure losses against, including:

- Property losses
- Liability losses

Insurance coverage is a must for snow contractors. It protects your company and your customers.

- Health insurance
- Life insurance

As a snow and ice management business grows, it will have to deal with all four. Property losses result from fire, accident, theft, or other identifiable perils. Liability losses result from property damage or injuries suffered by others for which the policyholder is held responsible. Other types of insurance we must deal with include workers' compensation, unemployment compensation, and Social Security.

Workers' compensation provides compensation for workers injured on the job. This includes payment of wages, medical care, job placement and, in some cases, vocational rehabilitation.

Unemployment compensation provides financial benefits, job counseling, and placement services for persons who become unemployed through no fault of their own. Social Security provides retirement benefits, life insurance, health insurance, and disability income insurance for the general population.

Snow and ice management contractors also have to deal with fire insurance on buildings used for operating a business, and automobile insurance on vehicles used in the operation of the business.

Other types of insurance that may come into play (both professionally and personally) include homeowner's insurance, business interruption insurance, and professional liability insurance (if you are dispensing advice in a professional context).

Workers' Compensation

As your snow and ice management business grows, you will begin to hire people to work with you. There are two types of people who will come to work for your company - subcontractors and full-time employees. We shall discuss working with subcontractors in greater detail in Chapter 10 *(See page 97)*, but in this chapter we shall focus on full-time employees.

Workers' compensation claims can put a company out of business if they are not taken seriously. While prevention is the preferred method of dealing with potential employee injuries, unfortunately such injuries do occur and must be dealt with.

Getting injured employees proper, timely medical care so they can return to work quickly is a key factor in keeping workers' compensation premiums within reason. Not paying attention to what is happening can prove disastrous in the form of increased premiums, decreased productivity in your work force, and a decided effect to your bottom line.

Additionally, putting one's head in the sand and assuming the insurance company will "take care of it" is not a sound business management practice. Make no mistake, the process should be viewed as injury management, since you can manage the process of getting an employee the proper care in an effort to return that employee to an active work capacity.

This in no way implies companies should take a callous attitude toward good medical attention for an injured employee. However, not paying attention to the proper aftercare may promote abuses to the system and, therefore, dramatically escalate the costs associated with such insurance coverage.

Proper Documentation. When an employee is injured in any fashion while on the job, a report of the circumstances surrounding the injury should be generated. Even minor injuries can balloon into major medical conditions if not treated properly. If an employee fails to report such an injury, it can cause procedural problems later on if no one at your company is aware of the incident.

It is a good idea to designate one person within your company to manage the documentation following the injury from initial report to final disposition. All communication, documentation, and information should flow through this one person so nothing is lost or overlooked. This includes regular communication with the injured employee, the insurance adjuster in charge of the case, the medical professional, and the company itself.

Your Medical Team. Having a panel of medical providers is essential to the entire workers' compensation process. Employees should be notified, when hired, that there is an approved panel of medical providers they are to see in the event of a work-related injury. While insurance carriers have a recommended panel of physicians, hospitals, and medical facilities they work with, the insured (your company) should be able to have input on what doctors and facilities are listed on the panel.

These providers have the same goal as the company – that is to be certain the employee receives the proper care and treatment, and returns to work as soon as possible after recovery.

Should the work-related injury require a hospital stay or medical attention, the person in charge of injury management has the right to, and should, accompany the employee to the doctor's office, hospital, or medical facility. Some companies insist the injury manager be with the injured employee as soon as possible after the

WORKERS' COMPENSATION AND HIPAA

The HIPAA Privacy Rule, enacted in 2003, recognized the legitimate need of insurers and other entities involved in the workers' compensation systems, to have access to individuals' health information as authorized by state or other laws.

Source: U.S. Department of Health and Human Services

injury occurs and then stay with the injured employee through the first hours of treatment, whether it is at a doctor's office, the hospital or the physical rehabilitation center.

Oftentimes, the injury is not an emergency and can be treated in a doctor's office or urgent care center instead of a hospital, thus saving considerable dollars for the company or the insurance carrier.

Once at the doctor's office or hospital outpatient facility, if further medical treatment is deemed necessary, the injured employee and the injury manager can be transported to the recommended facility. The injury manager should then communicate with the doctor in charge to ascertain what additional immediate treatment is needed (if necessary), as well as what aftercare may be required.

Aiding Recovery. Unfortunately, some injured employees may not adhere to the prescribed follow-up procedures and/or appointments, thus causing inadequate or insufficient attention to proper ongoing care. The injury manager should stay on top of the situation by communicating with the injured employee on a regu-

Employees returning from injuries can perform light-duty tasks such as basic vehicle maintenance.

lar basis. In some cases it is advisable for the injury manager to accompany the employee to the aftercare doctor visits to ensure the employee is adhering to the proper care program.

Keep in mind you are, essentially, footing the bill for the employee's care through your insurance premiums, thus necessitating your participation in getting the employee back to health and able to return to work. At these aftercare visits, the injury manager has the right to speak with the physician to ascertain what is needed to assist with the recuperation process.

Often, the case manager from the insurance carrier will also attend these appointments to keep accurate records of the entire process. Good communication is a key element in getting the employee back to work as soon as possible.

Getting Back To Work. Part of the injury management process is to have a light-duty work program in place so the injured employee can go back to work, earn a paycheck, and feel productive. Additionally, during this time of light-duty work, the injury manager can oversee the remainder of the recuperation process. Sometimes this light-duty work schedule must be tailored to the individual employee's needs.

It is common knowledge that performing light-duty work often accelerates the desire to return to work and positively affects the recuperation process. Unfortunately, in the middle of winter, this can prove a daunting task if there is no other work than snow plowing duties. A larger company can often find light-duty work within its core business base, but if you are just plowing snow, it will be more difficult to achieve an adequate light-duty, back-to-work program.

A huge factor in getting an injured employee back to work is communication. Communicate with the employee immediately after the injury occurs. Communicate with the medical professional to determine a proper care and light-duty work regimen. Communicate with the insurance company's injury management specialist so they don't keep writing checks to the employee for an unlimited amount of time if, in fact, that employee is able and medically cleared to return to work.

Continue to communicate with the injured employee so he knows you care, and that the company is concerned that proper care is being provided, and so you can keep tabs on the progress of the recuperation process.

A successful businessperson will recognize the involvement of the company in the process is key to keeping workers' compensation insurance premiums down, and helping your employee return to health.

Comprehensive General Liability

Snow and ice management professionals, and all subcontractors should be covered for comprehensive general liability insurance claims. In a vast majority of the markets that receive snow, service vehicles must have insurance coverage to be driven on state and municipal roadways. However, this coverage isn't usually adequate to cover potential losses that can occur during plowing operations.

Commercial vehicle coverage will take care of claims associated with an accident that directly involves the driver and owner of the vehicle. If you slide through the plate glass window at the local Sears store in the mall, your vehicular insurance policy (with commercial coverage) will take care of that problem.

However, should an accident occur *after* you have left the site (a slip and fall, or one vehicle slides into another on some patchy snow that remains on the surface after you have completed plowing operations), you will need a comprehensive general liability policy to assist in defending and satisfying a claim that may arise.

Simply stating to your customer (verbally or in writing) that you are not responsible for any such claims is not enough to keep someone from filing suit against you, or naming your company as a party to such an action. While you may not be found liable, you still need an attorney to defend you. If your coverage includes it, the insurance company will pay to defend you if you are named as a party in such a suit. Too often, inexperienced snow and ice contractors will write into their contract language "our company is not responsible for any accidents that occur on your premises." Not only is this foolish, it is wrong. This isn't that different from putting a sign in the back window of your car that says, "I'm not responsible for accidents I might cause." It doesn't work.

If you created the situation through your inactivity or failure to adequately address it at the site, you are at fault, regardless of what you might think, say, or write. Unless the contract language specifically states the customer agrees to "defend and hold harmless the contractor" for such situations, you will be forced to accept at least some responsibility for your actions (or, inaction as the case may be).

Customers will regularly request they be listed as "additional insured" on the certificate of insurance you supply them. This is normal and customary, and by doing this, the customer is asking that you defend them in the unlikely event that your actions (or inactions) while performing your services are deemed negligent in some fashion.

You should carefully read exactly what is being requested, and having a customer listed as additional insured

SNOW FACT:
SNOW DENSITY
The mass of snow, per unit volume, which is equal to the water content of snow divided by its depth.

65

TOP SNOWSTORMS IN UNITED STATES' HISTORY

#1 BLIZZARD OF 1888

March 11-12, 1888

Unseasonable and devastating snowstorm from the Chesapeake Bay to Maine. The cities of Washington, Philadelphia, Boston, and New York City were paralyzed. This incredible "Nor'easter" dumped 50 inches of snow in Connecticut and Massachusetts, while New Jersey and the state of New York had 40 inches. Drifts of 40 to 50 feet high buried houses and trains. From Chesapeake Bay to Nantucket, 200 ships were sunk with 400 lives lost.

Source: National Weather Service (www.crh.noaa.gov)

is acceptable. However being, listed on your policy as "*named* additional insured" is very, very dangerous. This wording means that your carrier may be asked to participate in any suit that names the customer. When you hire a subcontractor, your company should be listed as "additional insured" on the subcontractors' policy. You should be sure to review the insurance documentation on a regular basis to be certain all is correct when it comes to how the language is written.

There have been instances where an unscrupulous subcontractor will forge a certificate of insurance, and fraudulently claim to have proper and adequate insurance coverage. Unfortunately, some companies have found this out the hard way. For example, you may learn when a claim is filed and your insurance carrier enjoins the subcontractor that the certificate you received is bogus. As a result, you are on the hook for the entire defense of the claim.

This doesn't make you look real good to your insurance carrier, as well your customers. Never accept a certificate of insurance from the subcontractor. It should come from their insurance agent. You want an original copy that is mailed from the agent, and an occasional phone call to the agent, to be certain the certificate is valid.

Finding An Insurance Broker

Because insurance is complicated matter, take care in selecting your broker. In searching for the right broker to handle all your insurance needs, interview them as you would any potential partner or employee in your business. The right insurance carrier can indeed be a partner in the success or failure of your business.

Network with snow and ice management professionals you respect, and ask them about their insurance coverages, their agent, and their satisfaction level with both. Such time-tested advice can be invaluable and most contractors are willing to help you avoid the mistakes they've made.

Bonding

Should you need to supply a bond for a quotation your company is preparing, your insurance agent can assist in securing this. Bonding a job is an insurance policy against failure to perform to the expectations of the contract or customer.

Be very, very careful if a performance bond is requested for a snow or ice management project. Performance bonds are subject to someone ascertaining if you have adequately performed your duties under the terms of the contract. Performance in snow and ice management is very subjective, and wide open to interpretation. Do not be afraid to cross out this kind of language in a contract that comes from a your customer. Remember, a contract is a negotiating tool, and not gospel.

I once received a contract from a customer – interestingly enough, from the property manager for a major insurance carrier – that specified a fidelity bond. In discussion with the representative for the insurance company's building manager, we were told "all vendors *must* supply a fidelity bond – no exceptions." After a fair amount of discussion, it was finally decided to have our insurance company interact with their risk manager. It was agreed that whatever was decided by them would be followed.

When our insurance company representative told their risk manager that we were plowing snow – the risk manager immediately stated "you guys don't need a fidelity bond," something this author already knew to be true. Fidelity bonds are for coverage for employees of the vendor who might steal things from offices inside the building. Our employees would never be in the building, and therefore, there was no need for a fidelity bond.

The customer was flabbergasted the fidelity bond was waived, and made it clear that this was the first time this had ever happened. You must be diligent when reviewing customer-supplied contracts to be certain that clauses and conditions that put your company at risk are not included in the final contract.

If you have a question or concern about a clause or condition in a contract, consult your insurance agent or lawyer. Having them review the document is a good idea.

Your insurance carrier and you are partners in business. They will protect you in a time of need, and you should endeavor to not assume risks that are not yours to deal with. For large and involved contracts, sometimes showing them to the insurance carrier can prevent considerable heartburn later on. They *are* on your side.

Working with snow and ice presents numerous risks for contractors and proper insurance coverage is a must.

Chapter Seven
PRICING SNOW WORK PROFITABLY

Chapter Highlights

- Pricing Models
- Seasonal Contracts
- Combination Pricing
- Pricing Variables

O ne of the most often asked questions regarding the snow and ice management business is how to properly determine pricing. Pricing your services to match your market, or circumstances, is critical to the success of your business.

When snow contractors get together pricing is usually a hot topic of conversation. Unfortunately, some companies think snow and ice management services are a necessary evil instead of a highly profitable revenue source. Contractors with this attitude are often pricing their service incorrectly or unprofitably.

There are countless ways to price snow and ice management services, but most are variations of one of four types of pricing structures:

- Per-Inch/ Per-Event
- Per-Push/Per-Occurrence
- Per-Hour/Per-Piece of Equipment (or, also called Per-Event)
- Seasonal Contracts

Pricing Models

No matter what method your company uses to generate revenue for snow and ice management services, pricing should be one of the last areas used by custom-

ers when comparing potential service providers. Unfortunately, too many contractors use pricing as their primary sales tool. This devalues the industry as a whole and creates a market that competes on price, not service.

Pricing services should be done in a manner that is consistent with your company's profit goals and the needs of your customers. Customers who attempt to drive the decision making process based on price alone are short-changing themselves and our industry. The recession that started in 2008 created significant downward pressure on pricing. However, while some cus-

Determining how to price your snow removal services is an important step in building your business.

There are several pricing methods for plowing including per-event, per-push, per-hour and seasonal.

tomers actually expected contractors to lower prices for their services to be competitive. The savvy contractors found ways to charge customers lower prices, while still achieving acceptable margins.

In these times, good contractors searched for ways to reduce their own overhead costs, rather than just "chop the price" for their snow and ice management services.

The use of business management tracking and financial software to eliminate back-end office staff, forcing manufacturers to improve service quality and finding ways to force acceptance of liability exposures back upon the customers, are a few examples of how contractors were able to keep margins at an acceptable level during this period.

Customers who call to request "hourly rates" for snow and ice management may not be asking the question the right way, or do not realize what information they are actually looking for. For example, if a potential customer calls asking for your hourly rate, you can answer them this way:

"We have two hourly rates – $256.00 per hour or $35 per hour."

When asked what the difference is between the two, explain it this way: "At $256 per hour, I'll send a loader with a 14-foot capture blade, and do the job in 15 minutes. At $35 per hour, I'll send 15 guys with shovels, and they will do the job in four hours."

Now ask yourself, "What does the customer really want to know?" In truth, the customer wants to know how much you will charge them for clearing their parking lot, not necessarily what the hourly rate is for your equipment. The point is that part of the "pricing question" is to properly ascertain what the customer seeks, and communicate your prices in a manner that makes sense to the customer.

Per-Inch or Per-Event Pricing. At one time, pricing on a per-inch basis was reserved for very large accounts located in areas where snowfall levels varied greatly. Universities and airports are prime examples where per-inch contracts were typically used. Customers who requested this pricing structure normally had snow management budgets in excess of $250,000, and often figured on spending several million dollars on snow and ice management in a given winter season. This is no longer the case.

East of the Mississippi, customers are requesting this pricing model more frequently, and are doing so for smaller accounts. Customers would like to project more accurate costs for snow and ice management services. Customers are also learning that per-hour pricing is hard to control and manage. Quoting prices for such accounts requires knowledge on a variety of issues, including:

- Accurate production times for all pieces of equipment and manpower that might be used.
- First-hand knowledge of the type of snowfalls that might occur at the site.
- Probable moisture content of the accumulated snowfalls.
- Prevailing wind direction of the probable snow event.

Pricing snow management on a per-inch (or per-event) basis requires a strong familiarity with the intricacies of the account and the performance expectations. Software estimating packages allow inexperienced plowers to quote accurate production times for various pieces of equipment and makes pricing easier to do.

Services are priced for the entire event based upon how much snow accumulates at a given location. Zero to 3 inches, 3 inches to 6 inches, 6 inches to 9 inches and so on are examples of how the pricing is structured. As such, the snow contractor needs to evaluate how

many times they will need to service a site during any given snow event where there is measurable accumulation. This would include return visits to keep drive lanes open and clear. Some include deicing visits as part of the pricing structures. However, if you can provide deicing services on a "per-application" basis, you can achieve much higher margins for this ancillary service.

It used to be per-inch pricing was not for the inexperienced plowing contractor owning only a few pieces of equipment. On very large accounts, airports immediately come to mind, these often require liquidated damages if the required equipment is not available during a snow event. This type of pricing structure is much more profitable than per-hour pricing, but not nearly as profitable as per-push pricing structures.

Per-Push Pricing. Per-push pricing is a structure in which you bill the customer a consistent rate each time you visit their facility to plow snow. Plowing contracts billed on a per-push basis should be the most profitable. Again, it used to be that pricing on a per-push basis required considerable expertise, since you had to know what your or the subcontractors' equipment production capabilities were to properly project revenue generation.

As previously noted, software estimating packages on the market today can accurately indicate production factors for almost every type of equipment. Four-to-one and five-to-one ratios are common when comparing revenues to costs. An experienced contractor using an estimating package can now project accurate per-push costs on sites as large as 150 acres of paved surface. Pricing projects on a per-push basis also al-

SNOW FACT:

Most snowflakes are less than a half-inch across; however, they can get as big as 2 inches across. In 1971 a snowflake measuring 8 inches by 12 inches was recorded in Bratsk, Siberia.

lows customers to know exactly what to expect to pay when services are performed.

It should be noted that responsible contractors have a clause in their per-push contracts that allow for additional charges in the event snow accumulation exceeds a certain amount by the time they arrive on site. Additionally, if the contractor has to plow a particular site three or four times during the snow event, the per-push contract allows them to charge for each visit. In these cases it is recommended the contract with the customer allow the contractor to make the decision as to when to plow and apply deicing material.

There should also be a clause that advises customers "plowing and/or salting may not reduce the lot to bare pavement, and snow or ice accumulations are naturally occurring events, and the customer agrees to defend and hold harmless the contractor for trespass that may arise as a result of these naturally occurring events."

Per-Hour Pricing. Pricing per-hour is a structure in which customers are billed based on the time you spend at their facility. Pricing per-hour or per-truck is

Per-push pricing allows companies to bill customers at a consistent rate each time you visit the account.

the easiest way to avoid learning about snowplowing as a business. Such pricing methods allow the "guy next door" to get into the plowing business with little, if any, experience. Errors in judgment are paid for by the customer and not the contractor.

The margins are generally much lower, as contractors have a tendency to price to compete with the contractors who are also pricing by the hour. Growth patterns are accelerated, although at decidedly lower profit margins than are obtained by alternative pricing methods, because there isn't the need to visit every site prior to adding it to the service schedule.

Per-hour, per-piece of equipment pricing is the least profitable method of clearing sites. No matter the geographic market you serve, it is incredibly difficult to achieve anything above 30% gross margins when doing work "by the hour." Additionally, there is also little incentive for contractors to be efficient when performing work in this manner because the contractor does not care. Who cares how long it takes to clear the site? Certainly not the contractor.

In fact, there is huge incentive to be as inefficient as possible in these circumstances. It's also too easy to pad the customer's bill. Does the customer really know how long it took to clear the site, how much equipment was used on-site, or whether the guys worked a pattern to clear the site efficiently? There are customers and property managers who will argue the lowly snow contractor cannot possibly put one over on them, but the truth is more than a few contractors take advantage of this type of situation.

Due to scenarios such as the ones mentioned, there needs to be a tremendous element of trust between the customer and the contractor who prices their work on a per-hour model. Most snow contractors are honest and fair, but the unscrupulous contractors can add "ghosts" to a job site to increase revenues. This is because unsuspecting customers are not usually at the site at 3 a.m. to check how many pieces of equipment are being used. These contractors are caught eventually, which makes it that much harder for the honest contractor to generate a trusting relationship with the customer. Many national accounts require per-truck/per-hour pricing because it is easier for bidding purposes and, in those cases, it is often easier to take the account than to attempt to re-educate the customer.

Seasonal Contracts

In seasonal contracts, the customer is charged a consistent fee, typically by month, for the entire winter season regardless of the amount of snowfall. While these contracts typically include provisions for minimum and maximum snowfalls, they provide a consistent source of cash flow to the contractor.

To set seasonal rates, a contractor must determine the average number of snow events in their operating area. An event is defined as precipitation that requires a snow contractor to provide any service at a customer's location.

These services can include, but are not limited to plowing, deicing, sidewalk clearing and pre-salting. It's not an easy task to determine how many events your market area will receive each winter. History is necessary for such a determination, and a review of the weather occurrences for the past several years is required. Weather history information can be found on the Internet at various sites, including:

- www.millenniumweather.com
- www.weatherunderground.com
- www.nws.noaa.gov
- www.weathertap.com
- www.weatherbug.com
- www.accuweather.com (premium service)
- www.weather.com

Other ways contractors can find this information include the local library, local weather station meteorologists and local colleges or universities.

Once you have determined the average number of events that occur, you have to determine how much work will be required to serve the client during a snow event. While there is not a set of all-encompassing rules to determine this required level of work, past experience will be a tremendous asset. Previously mentioned software estimating packages can be helpful in determining pricing. Once pricing is determined and a value assigned to the total amount of work

necessary for the winter season, the contractor can recover overhead costs associated with the increased workload that winters normally bring.

Per-season pricing contracts are usually tied in with other services, such as landscape maintenance, parking lot sweeping, or a complete grounds maintenance service agreement. This allows for a year-round, all-inclusive contract outlining both summer and winter maintenance services. If you know the average number of times you plow in a given season, you can project how many times you will have to bring out the equipment.

Contractors who take on seasonal work for only one season can find it to be a disastrous situation. Someone almost always loses with a one-year contract. For example, if there are significant snow events, the contractor loses because of the escalated costs for the additional labor and equipment necessary to complete the work. If it snows

Snow contractors often reference weather websites such as www.weather.com to prepare for a major snow event.

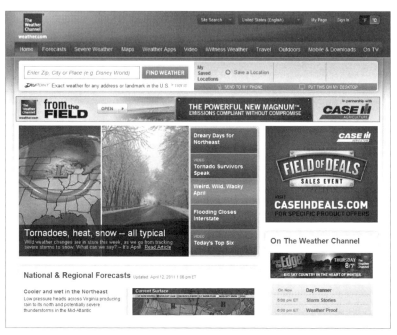

TOP SNOWSTORMS IN UNITED STATES' HISTORY

#2 ARMISTICE DAY STORM

Nov. 11-12, 1940

Mild weather ahead of an intense-low pressure system tracking from Kansas to western Wisconsin was quickly followed by a raging blizzard. Many people were caught off-guard by the severity of the storm and the plunging temperatures. Sixty degree temperatures during the morning of Nov. 11, were followed by single-digit readings by the morning of the 12th. These very cold temperatures and snow amounts were very unusual for this early in the season. Up to 26 inches of snow fell in Minnesota, while winds of 50 to 80 mph and heavy snows were common over parts of Wisconsin, South Dakota, Nebraska, Minnesota, Iowa, and Michigan. These winds were responsible for whipping up 20-foot drifts. A total of 144 deaths were blamed on the storm, most of which were duck hunters along the Mississippi River.

Source: National Weather Service (www.crh.noaa.gov)

less, the customer feels shorted because they perceive they did not get "their money's worth." With a three-year contract, both the contractor and the customer can assume the law of averages will make it even in the end.

In those years when there are above-average snowfall totals, the contractor might lose money on this particular account. However, if the other types of pricing models are used with other accounts, then the shortfall is generally made up due to the dramatically increased revenues generated by per-push and per-hour/per-truck pricing models employed with other clients.

Combination Pricing

If a snow contractor's business is priced solely on a per-push or per-hour/per-truck model, and there is a mild winter, revenues can drop below anticipated levels and cash flow problems can occur. Adding a mix of per-season customers can help contractors avoid these pitfalls.

For contractors in markets with snow removal revenues of less than $1 million a year, pricing structures normally fall in the per-push, per-hour/per-truck, or per-season categories. A combination of these pricing structures is preferred, allowing the contractor to maximize the benefits of each pricing structure, while minimizing the downside of a mild winter.

If a contractor can project the revenues they need to survive a mild winter season, these numbers should be readily available if you are doing your cost accounting properly, then securing enough per-season work allows the responsible contractor to guarantee adequate cash flow. Then, once this cash flow need is satisfied, remaining customers can be billed on a per-push, per-event, or per-hour basis.

Some contractors use a non-refundable retainer or minimum billing method to secure revenue prior to the start of the season. This has become an acceptable way of charging customers in the past 10 years. Obtaining several per-event accounts before the season

Pricing models for snow removal can vary depending on your geographic region.

starts helps the contractor's cash flow position before the first plow hits the driveway. Customers are then billed against the retainer for services rendered. If the number of events requiring service exceeds the amount of the retainer, the customer is invoiced for the additional services at an agreed-upon rate.

At one time, a large number of contractors in markets that received fewer snow events were using this pricing strategy as a way to generate significant revenues during slow snow seasons. More recently, contractors in high-volume snow markets are considering and implementing this type of revenue model because it is fiscally sound.

No matter what method or combination of methods your company uses, be aware that snowplowing is a viable profit center no matter how little or how much snow your geographical area receives. Making a profit at snow, and related snow services, is a mindset. As

long as it is thought of as a profit center like landscape installation, landscape maintenance, chemical lawn care, excavation, or power sweeping, money (good money) can be made from snow and ice management services.

Pricing Variables

Even with unique market and weather variables in your geographic region, marketing, pricing and sales operate much the same in Peoria as they do in Salt Lake City, even though Peoria receives wet and heavy snow, and Salt Lake City receives dry and fluffy snow. This is not to say there are no variables between markets. There certainly are and the variables can be significant from region-to-region.

I often hear arguments that one particular region of the country is mired in a per-hour/per-equipment pricing structure, and per-push pricing structures will not be accepted by customers. That notion is a falsehood.

Two things come to mind when I am confronted with that question. People said mass-producing motorcars was impossible because there could be no consistency in the parts manufacturing. Henry Ford proved this wrong. And, at one point, people said a single plowing contractor could not possibly manage to plow snow commercially in an area where he did not live. In both cases, challenges certainly were present, but those challenges were overcome with intelligent use of all the information available and by thinking outside the box. Yes, chances needed to be taken, but these chances were more than

"Look at the account and **tell the customer** the only way you can take on their business is at a per-push price."

dumb luck. They were well thought out and implemented with care.

For example, if you live and work in a market where snow management services are driven by the per-hour/per-equipment model and you want to go outside that paradigm to increase your margins on the work accomplished, you might approach it the following way.

Take on all the snowplowing work you can at the per-hour/per-equipment pricing structure. Once you have filled up your book with business and think you cannot take on any new clients, think twice before turning away the next call that comes into the office.

Look at the account and tell the customer the only way you can take on their business is at a per-push price. You

also will require a two-plowing deposit up front, which you will bill against as snow events occur. The customer may walk away and go elsewhere. However, you were prepared to walk away from the customer's business anyway, so what did you have to lose by proposing this?

Last-minute calls for snow and ice management quotes are usually the result of poor planning on the part of the customer. They may be desperate for a contractor for a variety of reasons – the previous contractor went out of business, they weren't happy with their current service provider or something else. It doesn't matter what the reason is and you really don't need the added work, but it gives you the chance to "step outside the box" and try something different.

It may take five of these calls to get one to consider this option, but eventually a prospective customer will take your quotation seriously. That will be followed by another and another. When that happens, you'll be able to drop a difficult and unreasonable customer from your list in favor of the customer who is willing to pay a portion of their bill up front.

There are contractors in markets who have taken this advice to heart and after a few years have only customers who pay per-push, and pay a portion of their invoice up front. Creating this pricing model within your company may take a while, but patience is a virtue that often handsomely rewards the diligent contractor.

If no one thought "outside the box" and tried to improve what they were doing, we'd all still be using two-piece phones in our homes and buying new sets of encyclopedias every few years to do research.

SNOW FACT:
The heaviest 24-hour snowfall on record in the continental United States happened in Silver Lake, Colo., when 76 inches fell April 14-15, 1921.

Chapter Eight

LAW, TAXES & CONTRACTS

Chapter Highlights

- Business Law
- Business Tax Fundamentals
- Creating Contracts
- Customer Supplied Contracts

The business world has certainly changed over the last 50 years. Gone is the handshake deal and your word is not good enough to get by when dealing with customers. Larger, national organizations that hire professionals to manage snow and ice for their sites simply will not do business without a written and fully executed contract in hand.

This is the way business is conducted today, but there is nothing to be afraid of when it comes to working with contracts. However, our customers hire lawyers to protect their interests, not your company's. In doing so, they will write contracts that favor their clients.

Most new snow and ice management contractors starting out, or only in business for a few years, do not have the luxury of full-time, in-house counsel to review every document that comes through the door. It is advisable to have outside counsel review contracts that seem over-protective of the customer at the potential expense of the contractor.

Commercial customers understand their vendors need to make a profit. After all, they are business people. They realize that without profit a company cannot survive. No large retail facility

Contracts are a necessary part of today's business world.

wants to beat up its vendor to the point where it is cheaper for that vendor to go out of business rather than continue to service the account. What could be worse than Home Depot learning its plowing contractor closed up shop just before a 12-inch snowfall arrived? Do not, for one minute, think this is what Home Depot, Wal-Mart, or Joe's Deli want to have happen.

However, these people also assume we are all professionals who know our costs and have quoted a viable price for the work specified in the contract. Contracts between customers and vendors are there to make certain all parties understand what is expected of

The world of business is governed by laws and regulations.

the other. In addition to spelling out what services are to be supplied and for what cost, assignation of potential liability is a very important part of the written agreement, or contract.

Laws protect everyone, not just the big companies. Imagine the chaos if we had no laws to guide us – no speed limits to control how we drive, no limitations on the medical profession, or no restrictions on taking things that don't belong to us.

The world of business is also governed by laws. The trend during the 1990s seemed to be that government was stepping in more often to govern the behavior of the business community. Thus, we have more laws and regulations regarding sexual harassment, proper treatment of those with disabilities, proper treatment of employees with regard to hiring and firing, working with subcontractors and enforcement of environmental and safety laws.

As the snow and ice management industry evolved, companies involved in providing snow and ice management services, both selling/supplying and buying/receiving, spent time de-

fining who was responsible for what. Property owners and managers wanted snow contractors to assume all liability exposure when accidents occurred on their sites. Snow contractors wanted to push the exposure back upon the property owner or manager. The tussle continues to this day as each party tries to determine what is "fair" with regard to such responsibility.

Business owners often want to set their own standards. Unfortunately, there are unscrupulous people who spoil it for the rest of us. You will hear small-business owners say, "This law doesn't apply to me," or "I won't follow that principle of law." Often, the person starting out in business will believe it is too expensive to follow the laws and they cannot make a profit in doing so. These laws are implemented for the good of all concerned and guide us in how we conduct our business.

In the snow and ice management industry, contractors often believe laws are stacked against them. At snow industry conventions, contractors often bemoan certain fees associated with being in business. It is unfair to expect those contractors who abide by the laws to shoulder the added expenses associated with doing so.

Business Law

Business law refers to rules, statutes, codes and regulations established to provide a legal framework within which business may be conducted, and that are enforceable by court action. Business owners need to have a working knowledge of the laws regarding sales, contracts, liability, taxes and bankruptcy.

"These laws are implemented for the **good of all concerned** and guide us in how we conduct our business."

Statutory Law. Statutory law (written law) includes state and federal constitutions, legislative enactments, treaties of the federal government, and ordinances. You can read these laws, but it is often difficult to determine their meaning. This is why we have so many lawyers, courts and others involved in the legal profession.

Common Law. Common law is the body of law that comes from decisions handed down by judges in previous court cases. Common law is often referred to as "unwritten law" because it doesn't appear in any legislative enactments or treaties. Such decisions are called precedents, and they guide judges in handling of new court cases.

Tort Law. Tort law relates to wrongful conduct that causes injury to another person's body, property or reputation.

These laws come into play in the event of a slip-and-fall incident. Liability laws fall under tort law, and hold a business liable for negligence in performance of its duties with regard to safety and execution. If you fail to adequately clear a property while plowing, you may be held liable for any injury that may occur as a result of your negligence in adequately clearing the lot.

Business Tax Fundamentals

Tax laws can be very daunting for business owners in all industries. Often, tax issues lead to spirited discussions as to the merits of any one particular tax. Let's define the major taxes levied upon individuals and businesses:

- Income taxes are taxes paid on the income received by businesses and individuals. Income taxes are the largest source of revenue received by the United States government.
- Property taxes are taxes paid on real and personal property. Real property is real estate owned by individuals and businesses. Personal property is a broader category that includes any movable property and tangible goods. These taxes are based on their assessed value.
- Sales taxes are paid on merchandise when it is sold at the retail level.

The taxes snow and ice management professionals need to be the most concerned with are income taxes and the related taxes for employees and sales taxes. We must pay income taxes on the profit we derive from the fruits of our labors. Some states require companies collect a sales tax on the services provided. These taxes must be paid to the corresponding state at the appropriate time.

Snow and ice contractors should consult a tax professional with any questions regarding what taxes do or do not apply to them or their businesses.

Creating Contracts

A contract is a legally enforceable agreement between two or more parties. A contract is legally binding if the following conditions are met:

- An offer is made.
- There is voluntary acceptance of the offer.
- Both parties must give consideration (something of value, such as money) in exchange for services rendered.
- Both parties of the contract must be competent.
- The contract must be legal (example: gambling losses are not legally collectable).
- The contract must be in proper form (example: an agreement for the sale of goods or services worth $500 or more must be in writing).

A breach of contract occurs when one party violates the contract. Should a breach of contract occur, the following can happen:

- The person who violated the contract may be forced to live up to the agreement.
- Payment of damages may be awarded to a person who is injured by the breach.
- Restitution can occur whereby if the person fails to live up to their end of the contract, the other party could agree to drop the matter, and wouldn't have to live up to his part of the agreement either.

Snowplowing contracts vary widely from company to company and from region to region. However, there are consistent elements that should be included in a standard snowplowing contract:

Headnote. Each contract should feature your company logo and contact information at the top of the first page. This communicates clearly who the quoting company is and how customers can contact the firm. A professional logo and overall quality design of the document creates a positive image for the company.

Submitted To. A brief description or representation of the company, the main contact and who is receiving the contract at the company.

Description of Services. Describes what the customer is receiving as part of the contract. This area can be quite simple and to the point. A brief description of what is to take place, when and how much this service will cost are pertinent details the customer will want to know up front.

Services Not Included. Some contractors also list related services that are not included as part of the contract, and how those "other" services will be billed. For example, in parts of the country where heavy snows are not routine, a contractor might charge an extra fee when physical removal of snow with heavy equipment is required. This section provides an area for the contractor to clearly communicate these special services to the customer.

Insurance. Some contractors include language describing the insurance they hold and if the company is bonded.

Terms and Conditions. The terms and conditions of a contract is where the meat of the document should lie. This is where you detail items that deviate from the norm, or any special conditions the customer needs to be aware of prior to agreeing to have you service their property.

A professionally printed, three-part contract with these terms and conditions preprinted on the rear is an excellent way to make sure all parties know what is expected during the term of the contract. However, in today's age of laser and inkjet printers, sometimes a three-part contract won't work. As such, a two- page document is necessary – page one for the quotation and page two for the terms and conditions.

At the bottom of the proposal's front page, where the customer signs to accept the contract, include language that specifically states the customer has read and accepted the stated terms of the contract.

In the terms-and-conditions portion of your contract you should detail *all* conditions under which you are accepting responsibility for the services you are to provide. While you cannot abdicate all responsibility for anything that might occur during or after a snow or ice event, it is possible to shift some of the liability for accidents to another party (ostensibly, the owner of the site you are servicing). Language to that effect should be contained in the terms and conditions of the contract.

Remember, it is not sufficient to simply state, "We are not responsible for this or that." You need to make it clear the customer agrees to "defend and hold harmless the contractor" for any trespass you wish them to accept responsibility for. By signing, they then agree with you shifting the responsibility to them, and that you have protected yourself.

An example of contract language that can help snow professionals limit their liability comes from Scott McEachern, program manager at Greensure, in Oshawa, Ontario, Canada. According to McEachern, contracts should be worded in a way that financial responsibility is *not* assumed when any of these three

Contracts protect your company if a question or dispute arises with a customer.

following concerns are present:

- In any and all cases in which the contractor is restricted by the property owner from sanding or salting at the contractor's own discretion.
- In any and all cases in which the damages or injuries to persons or property, or claims, actions, obligations, liabilities, costs, expenses and fees arise as the result of incidents occurring on areas of the property not serviced by the contractor.
- In any and all cases in which the damages or injuries to persons or property, or claims, actions, obligations, liabilities, costs, expenses and fees arise as the result of incidents occurring during times in which the contractor is restricted from accessing the property to perform his or her duties.

While a contract is a legal document that binds both parties to the agreement, you should be aware the court system generally treats residents differently than commercial, industrial and retail customers when interpreting the agreement in the event of a dispute. Residents are considered not as savvy about legal matters as those who own and operate businesses. It is important that residential customers fully understand what they are signing. Placing the text about having read and accepted the terms and conditions in **bold black** type demonstrates that you did not attempt to hide anything from an unsuspecting residential customer.

In contrast, commercial customers are often considered to be savvy about signing contracts, as this is part of the ongoing business duties of any owner or company manager.

Customer-Supplied Contracts

As your company grows and begins serving larger, more sophisticated clients, you will find many of the contracts will be written by your client. These contracts usually are written by the customer's legal department and often customers will ask that you sign their version.

This happens more often than one might think. In fact, it is standard practice with large clients who have multiple-locations that need to be serviced. Many times, these sites are located across the country and are serviced by independent snow and ice contractors of various sizes and types.

Read customer-supplied contracts closely. Remember, these contracts were written by the company's lawyer and are geared to protect the customer, not you. You should not refuse to sign a customer-supplied contract, but simply use normal business caution and perform your due diligence before signing. Consult an attorney if you have serious concerns about a customer-supplied contract or any contract or legal document.

After you have carefully read the contract, the decision of whether to sign it becomes a business decision. Additionally, these contracts should not be viewed as a take-it-or-leave-it proposition. Negotiations are part of the overall contract process. Often, the customer would like you to believe the only decision to be made is whether or not to accept the terms and conditions in the agreement. Do not be afraid to eliminate portions of the written contract you find to be overbearing or just plain incorrect. Rarely will the customer refuse to negotiate. Your approach should be to come with an open mind and willingness to give and take during the negotiating process. More often than not, you can eliminate areas of the contract language without fear of retribution.

If you have received a contract from the customer, it is likely you are their vendor of choice. Negotiating specifics of the contract is an accepted practice. For example, a client received a contract from a prospective customer. After reviewing the language, he decided six or seven items were unacceptable. The contractor "red lined" out the items and then submitted the revised version for consideration, but not without some considerable trepidation.

To his surprise, the customer came back and said, "We can accept all the changes except for one specific sentence." The contractor was amazed at how easy it went and obtained a contract that was more reasonable in scope and nature. This, too, is a business decision that must be made on the part of your potential customer. However, one might be surprised as to just how much negotiation can take place. As the old saying goes, "It never hurts to ask." Using a written contract is good business and good practice. It puts everyone on the same page.

NOTES:

Chapter Nine
FINANCIAL MANAGEMENT

Chapter Highlights

- Strategic Planning
- Business Planning
- Creating and Following A Budget
- Understanding Financial Statements
- Managing Cash Flow

Good financial management is essential to the success of any business entity. To be successful in your quest to grow your snow and ice management company, a working knowledge of financial management is imperative.

A basic knowledge of budgeting, planning, cash flow and the ability to read a financial statement go a long way to enjoying success in the snow and ice management industry. Often, contractors are concerned about getting the work done and don't pay attention to their finances. If you can manage finances and operate by the numbers, you can easily determine if you are operating profitably. And make no mistake about it, profitability is a must to grow a business.

Additionally, and maybe more importantly, you need adequate cash flow to stay afloat. Most companies, big or small, snow business or not, don't fail for lack of profit – they fail due to lack of *cash*. Cash is king and always will be. Financially solid companies do rolling cash projections to know where they stand at all times.

Strategic plans, business plans, financing plans, cash projections, forecasts and budgets all intend to do the same thing – assist in planning and controlling your business. Basically, these are all terms for the same thing. Unfortunately, the business and financial communities – even the government – tend to misuse them. Most often they will select a term they are familiar with and then apply it across the board. This can lead to catastrophic events that cannot be fully controlled.

In this chapter, we will define the various terms used in the business and financial communities and address those that apply to what growing snow and ice management companies need to know.

Strategic Planning

A strategic plan is a company's long-term plan that typically looks five or more years into the future. This type of plan lays out a company's long-term objectives in narrative fashion, covering all major activities including markets, services, finances, human resources and (in our case) subcontractor use. This narrative might be supported by detailed financial forecasts including balance sheets, income statements and cash-flow schedules for each year of the

Strategic and business planning involves setting the direction for your company and employees.

plan. We won't discuss strategic plans in depth since they are usually reserved for large, multi-million dollar companies that have to deal with a board of directors or stockholders. Nonetheless, it is never a bad idea for an owner of a snow and ice management company to think long-term and set business goals that are five or more years out.

Business Planning

A business plan covers a shorter term than the strategic plan, usually one or two years. Often the business plan will also function as a basis for financing plans that are given to the bank when applying for loans. The owner should use this plan as the working tool to develop tactical plans for monitoring the company's performance and development of its growth objectives.

If the plan is used internally, it may be short in nature but succinct in its level of detail. However, if used as a financial tool, it can be quite lengthy. In this instance, the business plan would include background information on the company or the company founder if a new

entity, a marketing plan, operating plan, staffing plan and financial projections. Parts of the business plan include:

The Mission Statement. The first section of a business plan is the company mission statement. This may seem simple because, after all, the immediate plan of any company is to make money. However, most companies have far more specific, or broader, missions than merely making money.

If you have not addressed the subject of the company's direction with your business partner(s), spouse or family, this is the time to do so. Without a definable purpose, goal or mission, a business of any size can tend to wander aimlessly through the daily maze of tasks, moving back and forth with the winds of change in handling customer demands, financial resources and even the whims of the owner.

A small startup company's mission statement may incorporate some strategic goals, but it is primarily aimed at achieving specific objectives in year one and perhaps year two. The mission statement should be geared more toward the details involved in operations, not financials.

Company History. Another part of the business plan is providing background or history of the company or owner. This section should remain constant from year-to-year and although there is nothing in this section that directly influences the character or the attainability of the plan itself, the history is a constant reminder of both wise and unwise decisions made in prior years. Having the company's history readily available also has a psychological influence on employees and others who may read it. Remembering the past brings the present into perspective.

Also, and this point is seldom understood in today's fast-paced, "need-it-now" business environment, the past usually repeats itself. Keep in mind, one definition of insanity is doing the same thing over and over again but expecting different results.

Capitalization. The capitalization of a company or the structure of a company's debt and equity (net worth), is more than just an accounting function – it is the foundation upon which assets are built. A company's capitalization consists of two segments: debt capital and equity capital.

Debt Capital. Debt capital represents all outstanding short- and long-term loans, plus mortgages. Total debt capital reflects a company's debt and how much cash must be paid out to creditors in current and future years.

Equity Capital. Equity capital represents ownership in the company and has two segments: 1) The amounts invested

in the company by owners; and 2) Earnings from prior years that have been retained in the company's coffers. The relationship between debt capital and equity capital, known as debt-to-equity ratio, is a key measure of the solvency of a business and its ability to meet its financial obligations and acquire assets. The higher the debt-to-equity ratio, the less money that is available for operating expenses and growing the company.

Additional Sections. Other important sections of the business plan can include information on your perceived competition, the service offerings you provide, pricing structures, advertising and sales promotional ideas and tactics, customer-service programs and collection policies. Another important section would be your management organization chart. This can be as small as you and your assistant, or it can be as large and detailed as needed to encompass your office staffing, sales team, field managers, departmental managers and frontline production workers.

Financing Plan. A financing plan is used almost exclusively to project your company's ability to secure and to repay loans. Designed primarily for large contractors, those projecting several million dollars in revenue and who have strong financial backgrounds, financing plans have very little narrative and focus heavily on financial projections. Such projections will cover three to five years, and are prepared using the same forecasting techniques as strategic and business plans. A financing plan is typically prepared by an internal accountant or controller.

Forecasting. Forecasting isn't necessarily a plan like others described in this

SNOW FACT:

About 70% of winter storm-related deaths occur in automobiles. The rest are primarily due to heart attacks from overexertion, such as shoveling heavy snow or from hypothermia caused by overexposure to the cold.

chapter, but instead a technique for projecting the most probable operating results, in a dollars format, over future periods of time. These periods can be months, quarters or years and forecasts usually include three basic documents:

- Pro-forma balance sheets
- Pro-forma income statements
- Cash flow schedules

Forecasts also form the basis for quantifying business valuations when buying and selling a business or shares in a business, for estate tax purposes and other situations that require a dollar value to be placed on a business entity. For purposes of this book, we shall not spend a lot of time on forecasting, with the exception of cash projections – which are key to the success of any business venture.

Cash Flow. Cash flow, for our purposes here, is the movement of cash in or out of the business. It is the cycle of cash inflow and outflow that determine your business' solvency. "Cash" is used here in the broader sense of the term, where it includes bank deposits. It is usually measured during a specific, finite period of time. Quarterly is the most common time frame for snow and ice management companies and extends throughout the year. Measurement of cash flow can be used for calculating parameters that give information on the companies' value and financial situation.

Cash flow can be used for calculating the following financial parameters:

- To determine a project's rate of return or value. The timing of cash flows in and out of projects are used as inputs in financial models

such as internal rate of return and net present value.
- To determine problems with a company's liquidity. Being profitable does not necessarily mean being liquid. A company can fail because of a shortage of cash, even while profitable.
- As an alternate measure of a company's profits when it is believed that accrual accounting concepts do not represent economic realities. For example, a company may be quite profitable but generating little operational cash.
- Cash flow can be used to evaluate the "quality" of income generated by accrual accounting.

Cash-flow analysis is the study of the cycle of your company's cash inflow and outflow, with the purpose of maintaining an adequate cash flow for your business. It also provides the basis for cash-flow management to pay your bills.

Cash-flow analysis involves examining the components of your business that affect cash flow, such as accounts receivable, inventory, accounts payable and credit terms. By performing a cash-flow analysis of these components, you'll be able to more easily identify problems and find ways to improve your cash flow.

A quick and easy way to perform a cash-flow analysis is to compare the total unpaid payables, subcontractors owed money for services rendered, outsourced repairs for equipment or salt purchases not yet paid for, with the total amounts of revenues due you at the end of each billing cycle. If the total unpaid payables are greater than the total revenues due to you, you'll need to spend more cash than you receive in the next month, indicating a potential cash-flow problem.

Operating Budget. An operating budget covers the shortest time period in any business plan. The operating budget may be for a week, a month or a year but seldom extends beyond 12 months. A budget is a formal, numerical expression of how a business or a division of a business should operate over a specific time period. Budgets are extremely useful as management tools for planning and controlling segments of a business, such as snowplowing or ice removal.

These segments can include departments, cost centers, profit centers or functions. The control feature is the most useful aspect of budgeting. Actual results for the week, month, quarter and year can be quantifiably compared to expected or projected performance, allowing you to take immediate, corrective action, if necessary. While some individuals try to incorporate sales projections into the budgetary process, cost budgets appear to be the most useful in smaller-sized companies. Cost budgets are often called operating budgets to differentiate them from more comprehensive planning

documents and we will review this in greater detail.

Fixed vs. Variable. Budget costs are either fixed or variable according to how they change in relation to other areas of the business. Operating costs are either fixed or variable in most instances, but sometimes they can be semi-variable. A fixed cost is one that is independent of other costs, sales or asset changes. In other words, fixed costs will be incurred independently of other transactions affecting the operation.

An example of a fixed cost is rent since it is held constant by a lease agreement. Another example is comprehensive general liability insurance premiums that are usually constant for a one-year period. In the absence of a major change, such as the renegotiations of a lease, actual and budgeted fixed costs should be identical each month.

On the other hand, variable costs are incurred in direct relation to other costs, sales or asset changes. For example, your electricity bill goes up or down in direct relation to how much is used each month. Travel expenses are directly related to the number of trips taken in a month or quarter. Fuel expenses are

The goal of a budget is to be realistic and attainable within the planned time.

Vehicles are a fixed or long-term asset of your company.

directly related to the number of snow events that occur and trucks you have on the road servicing customers during the winter season.

Controllable vs. Non-Controllable Costs. Controllable costs are those that you, as the owner or manager, can control by choosing to either incur or not incur. This would be a decision that often is made without regard for short-term sales volume or production level fluctuations. Some examples of controllable costs would be:

- The number of people you have on your payroll.
- The amount spent on travel and entertainment.
- The use of company vehicles.
- The decision to purchase office supplies or office equipment such as computers, cell phones, iPads, etc.

These costs may be fixed or variable over the budget period and you can de-

cide to incur them or not.

Some costs are beyond the scope of short-term decisions and these are non-controllable costs. They might include building rent, payroll taxes, property taxes, depreciation and local phone charges.

Generally, non-controllable costs remain fixed for a period of time. However, over two or three years, many can be renegotiated to lower levels or eliminated completely.

Creating and Following A Budget

The goal of a budget is to be realistic and attainable within the planned time frame. To be an effective management tool, operating budgets must not represent pie-in-the-sky dreams. They cannot be contrived targets aimed at motivating employees to "stretch" sales goals or production activity since today's employees do not respond to incentive goals unless they perceive them to be achievable. It should be noted that one of the main objectives of having operating budgets (if you have employees) is to get your people to improve their work performance.

There is a tendency for owners who have employees who are privy to the budget numbers to use them as weapons to prevent others from taking certain actions. For example, if your operations manager requests a new computer and you say, "No, the budget won't allow it," you, in reality, might not want to buy new computers for some other reason only you know about.

I would caution business owners from doing this. Using budgets in this fashion destroys their effectiveness since it doesn't take an astute employee long to see through this tactic. When this happens, budgets lose their credibility and they will no longer be an effective management tool.

Budgets should be built from the bottom up, not from the top down. Once you involve your employees in the budget process, make sure to give them a timetable to follow that allows for you to review with them their budget projections. Ultimately, as the owner, you will approve the budget, but building it involves good communication and compromise from everyone involved.

Understanding Financial Statements

The two basic financial statements applicable to every company, corporation or sole proprietorship, are the balance sheet and the income statement.

Balance Sheet. Balance sheets are statements in time. They show the value in dollars of the assets and liabilities of a company at a given time. These values will vary from day to day and the values of the balance sheet are valid only for that specific point in time.

The "Asset Side." The left side of the balance sheet shows the dollar value of all the company's assets at a given time. Current assets are listed first, followed by fixed, or long-term, assets and then intangible assets. Current assets can include:

- Cash
- Receivables
- Inventory
- Prepaid expenses

Fixed, or long-term, assets can include:

- Land
- Buildings
- Vehicles
- Equipment
- Long-term investments

Buildings, vehicles and equipment can be depreciated. This means their cost is written off over the length of their useful lives. The amount of depreciation to date is offset against the cost of these assets in a balance sheet. Land and long-term investments cannot be depreciated.

Intangible assets are comprised of various expenditures for organizational expenses, patents, trademarks, copyrights, etc. In other words, this would be any expenditure that adds value to the company, but it cannot be touched or held such as vehicles or equipment.

The sum of current assets, fixed and intangible, represents the total financial assets a company has at its disposal to generate income or to generate cash.

The "Liability Side." The "liability side" of a balance sheet has nothing

SNOW FACT:
Winter Weather Advisory. Issued for accumulations of snow, freezing rain, freezing drizzle and sleet which will cause significant inconvenience and moderately dangerous conditions.

to do with "liabilities" associated with slip-and-fall lawsuits – the two are unrelated. The right side of the balance sheet shows the dollar value of all obligations or liabilities of the business. These amounts represent amounts owed to all creditors. These creditors include: banks, suppliers and government agencies. The total amount owed to these creditors is called debt.

Examples of liabilities on a balance sheet can include:

- Bank loans
- The portion of any long-term debt that is due within one year.
- Accounts payable
- Accrued expenses, that represent expenses incurred, but not yet paid for interest expense, wages, payroll taxes, income and sales taxes, etc.

Long-term liabilities can include:

- Long-term notes
- Mortgages
- Corporate bonds

The sum of current liabilities and long-term liabilities represents the total amount owed by the company. When total liabilities are subtracted from total assets, the remainder is called net worth or owner's equity.

The Income Statement. The income statement, at times referred to as the "profit-and-loss statement," shows a series of transactions that have occurred over a period of time – specifically, the period of time between two balance sheet dates. This may be a month, a quarter, a year, or any other period of time. However, the income statement typically does not stretch beyond one year. All transactions that have occurred during this period show up in the income statement – sales, costs of materials and labor, operating expenses, taxes and interest paid on loans.

Cash Flow Statement. There is a third report called a "statement of cash flow." This is not considered a financial statement in an accounting sense, but it can tell more about a company's performance than either a balance sheet or an income statement. A properly prepared cash flow statement shows a company's total cash receipts and cash disbursements during the period of time covered by the income statement.

The relationships among the numbers in these three reports can enable a company owner to analyze the complete financial performance of the business. These relationships are usually expressed in ratios. Ratios that describe the various relationships among accounts in the balance sheet and income statement are the primary analytic tools of financial analysis. They are used to judge the financial health of a business, to project future earnings and cash flows and to value a business. Banks use these ratios to determine the viability of a business before lending to it.

The current ratio, current assets divided by current liabilities, is a measure of liquidity. One widely accepted standard is that a current ratio of 2:1 indicates a company has sufficient liquid assets to meet its current obligations. This standard is usually used when applied to larger companies that must follow generally accepted accounting principles. However, small businesses may have hybrid accounting systems that do not conform to these accepted accounting principles. In these cases, some banking

	Oct-01	Nov-01	Dec-01	Jan-02	Feb-02	Mar-02	Apr-02	May-02	Jun-02	Jul-02	Aug-02	Sep-02	Totals
Snowplowing Service Income	75,000	115,000	115,000	215,000	275,000	215,000	75,000	0	0	0	0	0	1,085,000
Gross Payroll	7,183	7,833	7,833	8,267	8,267	8,267	8,267	8,267	8,267	8,267	8,267	8,267	97,252
Gross Payroll													0
Subcontractor Labor	0	50,000	100,000	125,000	125,000	100,000	10,000	0	0	0	0	0	510,000
Office Supplies Expense	445	200	200	200	150	150	100	100	100	100	100	100	1,945
Miscellaneous Expense	250	50	50	50	50	50	50	50	50	50	50	50	800
Postage Expense	77	50	20	20	20	20	50	50	50	50	50	50	507
Printing Expense	14	25	15	10	10	10	15	20	20	20	20	20	199
Marketing Expense	250	250	250	250									1,000
Travel Expense	845	100					50	100	150	200	200	200	1,845
Memberships/Subscrpt/Lic							25	25	25	25	25	25	150
Vehicle Expense	155	150	150	150	150	150	150	100	100	100	100	100	1,555
Meals & Entertainment	155	25	25	25	25	25	50	100	100	100	100	100	830
Insurance (Group Health)	2,800	2,800	2,800	2,800	2,800	2,800	2,800	2,800	2,800	2,800	2,800	2,800	33,600
Insurance (Workers Comp)	3,300	3,300	3,300	3,300	3,300	3,300	3,300	3,300	3,300	3,300	3,300	3,300	39,600
Insurance (Bldg,Theft,Fire)	250	250	250	250	250	250	250	250	250	250	250	250	3,000
Insurance (Auto)	500	500	500	500	500	500	500	500	500	500	500	500	6,000
Rent Expense	200	200	200	200	200	200	200	200	200	200	200	200	2,400
Water & Sewer	100	100	100	100	100	100	100	100	100	100	100	100	1,200
Electric	20	20	10	10	10	10	20	20	20	20	20	20	200
Gas Heat Expense	20	20	20	20	20	20	20	0	0	0	0	0	140
Telephone Expense	378	25	25	25	25	25	25	25	25	25	25	25	653
Cable TV Expense	5	5	5	5	5	5	5	0	0	0	0	0	35
Vehicle Gasoline	693	350	400	450	450	450	450	400	400	400	400	400	5,243
Shop Expense	500	250	250	250	250	250	250	100	100	100	100	100	2,500
Computer Hardware/software	200	200	200	200	200	200	200	200	200	200	200	200	2,400
Payroll Tax - FICA	550	600	600	633	633	633	633	633	633	633	633	633	7,447
P/R - State Unemployment	269	247	247	856	856	247	247	247	247	247	247	247	4,204
P/R - Fed Unemployment	19	19	19	66	50	19	19	19	19	19	19	19	306
Employee Bonus Program													0
Temporary Employee Exp	949	949	949	949	949	949	949	949	949	949	949	949	11,388
Semiar & Training Expense									500	500	300	300	1,600
Simple IRA Matching Exp													0
Interest Expense	50												50
Total Expense	20,177	68,518	118,418	144,586	144,270	118,630	28,725	18,555	19,105	19,155	18,955	18,955	738,049
Net Profit or Loss	54,823	46,482	-3,418	70,414	130,730	96,370	46,275	-18,555	-19,105	-19,155	-18,955	-18,955	346,951

Cash flow management is a critical part of running your company and promoting growth.

institutions may require that the ratio be 3:1 or 4:1 to ensure prudent cash management and adequate liquidity are present.

Managing Cash Flow

Cash is king. We've all heard this statement and it is true. A company that has cash to pay its bills is a well-run company. Current accounting methods permit companies to show a profit on their financial statements even though they don't have enough cash to meet their bills.

There is nothing complicated about cash-flow management. In the simplest of terms, cash-flow management is a set of procedures aimed at maximizing the amount of cash available at any given time. We all practice cash-flow management in our daily lives by tracking the checks we write and the deposits we make. This way we can be certain we have enough cash to cover tomorrow's checks.

The same principle applies to cash-flow management in a business. You have to make sure the company has sufficient funds in its checking account to pay its bills tomorrow, the next day and the day after. That's all there is to cash-flow management.

Unfortunately, most companies do not strictly practice this principle. Credit is dangled in front of companies and consumers every day. We are encouraged to buy on credit and borrow to pay credit obligations. It often gets to the point where a business cannot generate enough cash to meet its obligations. Sometimes, it causes good, hard-working people to go out of business.

Scrutinize your company's expenditures closely, collect receivables aggressively and have cash on hand. Some companies take 3% of all money that comes in the door and put it away in a separate cash account that is used for unforeseen expenses. Ultimately, a successful contractor is separated from a failed one based on his awareness of his company's financial health.

Chapter Ten
WORKING WITH SUBCONTRACTORS

Chapter Highlights

- Advantages and Disadvantages
- What Is A Subcontractor?
- Basic Business Elements For Subcontractors
- Insurance and Subcontractors
- Finding Subcontractors
- Managing Subcontractors
- Subcontractor Agreements

Growth in snowplowing depends upon how much equipment and manpower you can muster to deal with the increased business. Additional equipment requires a substantial capital investment if you choose to own all of the necessary equipment. Finding quality, reliable employees is also a challenge.

Using subcontractors is one proven method to reduce your capital investment and still take on new business. Subcontractors supply both the manpower and equipment necessary to adequately service your customers during snow and ice events.

For those who are unfamiliar with how to go about finding, managing and retaining quality subcontractors, this chapter offers ideas and guidelines to assist you. The chapter also explores the legal issues surrounding subcontractors as they pertain to the responsibilities of both the contractor and subcontractor.

Advantages and Disadvantages

There are both advantages and disadvantages with using subcontractors for snow and ice management. One of the greatest advantages is that you won't have to invest substantial money in equipment since the subcontractor will use his or her own equipment.

In the eyes of the IRS and the state employment bureau, this is one of the tests used to determine if someone is a subcontractor instead of an employee. Using properly classified subcontractors eliminates the burden of keeping track of and paying the standard employment taxes associated with an employee. This gives your company greater flexibility to expand its residential and/or commercial coverage area.

Employers must withhold and pay federal payroll taxes for employees. This includes a 7.65% Social Security tax and a small federal unemployment tax. The employee's Social Security taxes and federal income taxes must be withheld from the employee's paychecks by the employer and paid to the IRS. For qualified subcontractors, these taxes do not have to be withheld and forwarded to the appropriate agencies. The subcontractor is responsible for the taxes and filings, thus lessening the burden on the contractor.

Employers in every state are required to contribute to the state unemployment insurance fund on behalf of employees. However, these contributions are *not* necessary when using subcontractors. Keep in mind that in some states even a self-employed individual plowing snow must pay workers' compensation. Since each state has its own rules regarding this matter, check with your respective state agency.

Sounds great, doesn't it? All these taxes don't need to be paid, all the paperwork associated with having employees is no longer necessary and you

> "The state may also **audit your records** to determine if you have properly classified these subcontractors."

are saving on accounting and bookkeeping costs.

However, it isn't quite as easy as declaring everyone whom you hire will now be "subcontractors." Subcontractors must be properly classified.

There are several qualification tests that must be satisfied to ensure various governmental agencies do not reclassify your subcontractor as an employee, thus negating all the positives described above.

The IRS may audit your company and closely review the subcontractors to ensure you are actually treating them as subcontractors.

The IRS wants to see as many workers as possible classified as employees and not as independent contractors. It wants to immediately collect taxes based on payroll withholding. If the IRS audits your company and determines you have

misclassified employees as independent contractors, it may impose substantial penalties and interest.

Additionally, the state may also audit your records to determine if you have properly classified these subcontractors. They, too, would like to classify all persons working for you as employees.

What Is A Subcontractor?

A few myths exist about who is and who is not a true subcontractor. The first one is that "saying a worker is a subcontractor will make it so." Don't get caught in this trap. The mere fact that you call a subcontractor a subcontractor will not by itself convince the IRS or any state agency.

Government auditors look at the substance of the relationship between the hiring company and the worker. Mere formalities are not enough. A well-drafted and truthful subcontractor agreement is helpful in the classification process, but it might not be enough. Auditors will make sure the parties involved actually behave the way the agreement indicates they should.

The second myth is that people who work for more than one company are considered subcontractors. If you have the right to control the worker on the job, he or she may be classified as your employee even though they work for others at the same time. It's possible for a worker to have more than one employer at the same time. However, if you specifically state that your subcontractor must work for your company and no others, then you will likely give an auditor just cause to classify your subcontractor as an employee.

The third myth is the assumption that part-time and short-term workers are subcontractors. Don't think because you're only hiring a person for a couple of weeks the person must be a subcontractor. If you have the specific right to control the person's actions, most government agencies will classify these people as employees.

You can legally use people with their own plow rigs as subcontractors, provided they behave in a fashion that keeps them independent from your snow and ice management company.

There is a common test for classifying subcontractors used by most governmental agencies to determine whether workers are employees or subcontractors. The IRS uses the "IRS 20 Factor Test" to explain the different parameters the agency has historically used to measure control over workers and determine whether an individual is a subcontractor or an employee. Over the years, the IRS has simplified and liberalized the test. As a result, not all of the factors are as important as they once were.

Basic Business Elements For Subcontractors

There are a number of different items to keep in mind if you decide to work with subcontractors or become a subcontractor yourself. Here are some basic business elements involved with subcontractors.

Making A Profit Or Loss. Subcontractors can earn a profit or take a loss as a result of the services being performed. Employees are typically paid for their time and labor, and have no liability for losses incurred by their employer.

Working On Specific Premises. Subcontractors usually are able to work in numerous places to perform their services. Employees work where their employers tell them to work, usually on the employer's premises.

Offering Services To The General Public. Subcontractors may and are encouraged to offer their services to the general public. Employees offer their services solely to their employers.

Right To Fire. Subcontractors can be terminated only for violating their contractual agreement with the company. Employees typically can be discharged by their employer at any time for any number of reasons.

Furnishing Tools and Materials. Subcontractors furnish their own equipment (trucks, plows, loaders, salt spreaders, deicing equipment, etc.). Employees are typically furnished with all of the tools and materials necessary to perform their jobs.

Method Of Payment. Subcontractors are generally paid a flat fee for a project. In the snow and ice management industry it is customary to pay subcontractors by the hour. And since this has been done across the country for many years, the practice has stood the test of time.

When a particular payment method has become general practice, the method-of-payment factor will not be given great weight. However, it is

Subcontractors play a significant role in delivering snow removal services.

TOP SNOWSTORMS IN UNITED STATES' HISTORY

#3 THE GREAT MIDWEST BLIZZARD

Jan. 26-27, 1967

One of the biggest snowstorms on record to strike the Midwest occurred just two days after an extremely rare January tornado outbreak in nearly the same area. An intense "Panhandle Hook" storm tracked from New Mexico northeast up the Ohio Valley. Central and northern Illinois, northern Indiana, southeast Iowa, lower Michigan, Missouri, and Kansas were hit hard by this blizzard. Kalamazoo, Mich., reported 28 inches of snow, Gary, Ind., 24 inches, and Chicago 23 inches with winds of 50 mph creating drifts of up to 15 feet. Seventy-six people died, most in the Chicago area.

Source: National Weather Service (www.crh.noaa.gov)

recommended subcontractors only be paid once per month since employees can be paid – usually by a unit of time – each week or every other week.

Subcontractors should invoice the company for work performed. If paid once per month, it cannot be erroneously assumed this person is an employee since federal labor laws specifically prohibit paying employees once a month.

Working For More Than One Firm. Subcontractors must have the ability to work for more than one client. This is a common mistake made by snowplowing companies when hiring subcontractors. Unlike the method-of-payment test, this test carries considerable weight when a governmental agency is determining whether a person is an independent contractor. Usually, an employee is expected to have some measure of loyalty, and employers can prevent employees from taking alternative or second jobs.

Continuing Relationships. Independent contractors are usually not contracted to work continuously for one company and this is typically true of subcontractors in the snow and ice removal industry. It is not reasonable to expect a subcontractor will be performing snow-related services in summer months. Employees normally have a continuing relationship with the employer.

Investing In Equipment/Facilities. Employees generally have no investment in equipment or facilities. Conversely speaking, the very nature of the services to be performed requires that the independent subcontractor have a significant investment in the equipment being used.

By making such a financial investment, the subcontractor risks losing the investment if the business is not profitable. Owning the equipment also implies that the subcontractor has the right to control its use.

Business Or Travel Expenses. Subcontractors typically pay their own expenses, such as gasoline, insurance, taxes, supplies and maintenance. On the other hand, employees' job-related business and travel expenses are paid by the employer.

Right To Quit. A subcontractor is legally obligated – by virtue of the written subcontract agreement – to complete the work he or she agreed to do. Employees may quit the job at any time without incurring any liability to the employer.

Instructions. Employers have the right to give their employees oral or written instructions that they must obey about when, where and how they are to work. Subcontractors need not comply with instructions on how to perform their services. Subcontractors decide how to do the work. As brokers of snow and ice removal services, we can and do give subcontractors instruction. However, those instructions are usually about the customer's expectations for a completed and acceptable product.

Sequence of Work. Employees may be required to perform services in the order or sequence set for them by the employer. Subcontractors generally are not instructed as to the sequence of events required to perform snow and ice management on specific sites. However, the customer may have a sequence that is required for adequate and acceptable performance standards.

For example, a customer may determine the front of a store should be plowed before the rear. This is not considered instruction or sequence on the part of the plowing company, as it is just working under the instructions given by the customer.

Training. Independent contractors generally receive little or no training from those that purchase their services. Employees are trained by their employers.

Services Performed Personally. Subcontractors are not generally required to render services personally. They can hire their own employees to do the work. Employees, however, cannot hire others to do their jobs for them.

Hiring Assistants. Subcontractors hire, supervise and pay their own assistants or employees. Employees hire, supervise and pay assistants only at the direction of the employer.

Set Working Hours. Employees ordinarily have set hours of work. Subcontractors generally set their own work hours. In the case of subcontractors, it is the customer who requires plowing be completed by a set time. We, as the broker, pass that information on to the subcontractor so everyone can meet the expectations of the customer.

Working Full Time. Subcontractors are free to work when and for whom they choose. They also have the right to work for more than one client at a time. However, employees are usually required to work full time for that one employer.

Oral Or Written Reports. Subcontractors are generally not required to submit regular reports. Instead, they are responsible only for end results. On the other hand, employees may be required to submit to the employer regular oral or written reports regarding the progress of their work.

Integration Into Business. Subcontractors' services are typically not molded into the hiring company's overall business as one integrated operation. Employees typically provide services that are an integral part of the employer's day-to-day operations.

A signed agreement between a contractor and a subcontractor keeps everyone on the same page regarding what is expected of each party.

**Independent
Contractors
Agreement
Snow Removal Services**

1406 West 21st Street, Erie, PA 16502-2203 • 455-1752
"Since 1978"

TERM: This agreement is made between Allin Companies, 1406 West 21st Street, Erie, PA 16502-2203 (called "The Company")

_____ DBA _____ (called "Contractor").
(Individual's Name)   (Company Name)

It will be effective commencing November 1, _____ and continuing until May 15, _____.

OBJECTIVE: The Company is in the business of providing snowplowing and removal, and similar services. Over the years it has developed a large number of accounts throughout Erie, PA and the surrounding suburban area and is continuously adding additional accounts for these services. These accounts are serviced primarily by The Company brokering out the work to independent contractors who provide their own equipment and perform the actual work. The intent of this agreement is to establish a working relationship between The Company and Contractor for this purpose. In order to establish and effectuate this relationship, the parties agree as follows:

1.   **CONTRACTOR SERVICES.** Contractor will be engaged by The Company as an independent contractor and agrees to provide services for The Company's accounts as directed by The Company during the period stated above. Contractor has the following equipment which will be provided, along with a qualified operator, at all times when requested by The Company:

EQUIPMENT

Year	Make	Model No., Transmission Color/Description	Size of Blade/Bucket	Rate per hour

2.   **PAYMENT.** The Company will pay the Contractor the rates shown above for the Contractor's services and/or equipment. Payment will be made about 10 days after receipt of invoice from the Contractor for services rendered in the prior month. It is recommended that invoices be submitted on (or about) the 1st day of each month. The Company requests only one invoice per month and under no circumstances will checks be issued more than once per month. No checks will be issued between the 1st of each month and the 10th of each month. In the event there are any damages done by the Contractor to the property of The Company's customers, or poor workmanship that has caused The Company to have to credit the customer's account, these damages will be paid for by the Contractor and shall be deducted from any amounts due the Contractor.

3.   **TRAVEL TIME.** Contractor will be paid for travel time between the various jobs that have been assigned to him. Agreed rate will begin at time Contractor leaves his home (in plowing vehicle and ready to plow) and will end at completion of the last job. Special arrangements will be made if Contractor lives outside Erie as travel time to first job will be paid for long distances. No travel time will be paid from the last job site on the assigned route. The Company does not pay for "breaktime" or "downtime". The Company does not pay for time required for fueling the Contractor's vehicle. If vehicle must be fueled prior to starting the assigned route, the start time will begin when Contractor leaves fueling point.

4.   **CONTRACTOR'S EXPENSES.** Contractor will provide all fuel for his equipment. All maintenance will be at Contractor's expense and his equipment is to be properly maintained to avoid breakdown during an urgent snow removal period. Contractor will not be paid when his equipment is broken down or inoperable.

5.   **SUBCONTRACTOR AND INSURANCE.** Contractor is engaged as an independent contractor and will not be considered an employee of The Company. Contractor will provide his own equipment as noted above, as well as any other tools or supplies which are necessary in order to provide the services which Contractor is engaged to provide. He must provide transportation for himself and his employees. He must have a telephone where he can be reached by The Company. He must provide any office or administrative services which he requires apart from The Company property and without cost to The Company. He must provide his own Worker's Compensation insurance for himself and for his employees. Contractor must provide auto liability insurance with recommended limits not less than $300,000. He will provide The Company with appropriate certificates of insurance. He will be responsible for payments to his employees for their work and for any required withholding, and pay overtime according to law. If, for some reason, additional insurance premiums and/or taxes are levied against The Company because of the services provided by Contractor, then those costs will be paid to The Company by Contractor. The Company may withhold such costs from any amounts that it has due to the Contractor from The Company. If there are not sufficient amounts owing Contractor by The Company to cover such costs at the time they become due by The Company, then Contractor shall repay them to The Company within 30 days from his receipt of notice thereof from The Company. If payment is not made, The Company may proceed to collect from the Contractor by legal process and in such case the Contractor shall also be responsible for The Company's reasonable costs and attorney's fees. In any event, failure to provide a certificate of insurance will result in 25% of amounts due to Contractor being retained by The Company to cover additional insurance premiums that may be levied against the Contractor by his insurance company in order to adequately insure that apparent uninsured

Worker Benefits. Subcontractors receive no standard workplace benefits. Employees usually receive benefits such as health insurance, sick leave, 401(k) programs, pensions and paid vacation.

Tax Treatment Of The Worker. Subcontractors pay their own taxes and are supplied with a 1099 form attesting to the fact that they have been paid monies without any taxes having been withheld. Employees usually have federal and state payroll taxes withheld by the employer and remitted to the government. Employees receive a W-2 at the end of the year attesting to these taxes and wages having been paid.

Intent Of The Hiring Firm And Worker. People who hire subcontractors normally intend to create an independent contractor-hiring firm relationship. People who hire employees normally intend to create the employer-employee relationship.

Industry Customs. Workers who are normally treated like subcontractors in the industry in which they work are likely to be independent contractors. This is the norm in the snow and ice management industry, so it is likely they are actually independent contractors. This is the same for employees. Workers who are normally treated like employees in a particular trade or industry, are likely to be employees.

The ABC Test. About half of U.S. states use a special statutory test – also called the ABC test – to determine if workers are subcontractors or employees for purposes of unemployment compensation.

This test focuses on a few factors, including whether the hiring firm controls the workers on the job, whether the worker is operating an independent business and where the work is performed. This test is much simpler than the IRS test, but it can be the most difficult to fulfill.

There are numerous other factors involved when setting up subcontractors to work with you in your endeavors to grow your business. This could be the subject of an entire book. If you would like to know more on the subject, I recommend reading "Hiring Independent Contractors: The Employer's Legal Guide" (Nolo Press) by attorney Stephen Fishman.

Insurance and Subcontractors

Subcontractors should have insurance. It is in your company's best interest to have insured subcontractors because, in the event an issue, they are responsible for any damages they create and for the portion of any liability that is assigned to your company.

Some states require vehicles driven on roads have a minimum amount of insurance coverage. In those states where it is not mandatory, you should insist on it for your own and others'

protection from catastrophic events.

If the subcontractor has employees, you should insist the subcontractor carries workers' compensation coverage at the state-mandatory minimums. Keep in mind, if a subcontractor's employees are hurt in some fashion on the job, and if the subcontractor is not carrying proper insurance coverage, you could be held liable for covering their injuries.

In addition to the above-referenced insurance coverage, the subcontractor should carry comprehensive general liability insurance. This insurance policy covers accidents that may happen after your subcontractor has left the job site. While you – the plowing contractor/broker of record – can take steps to limit the exposure of such liability in your contract language, you cannot totally divorce your company (or your subcontractors) from the possibility of someone filing suit against you.

Most insurance-savvy customers will insist on being named "additional insured" on your policy. This is normal and the plowing contractor should insist on the same "additional-insured" status on the subcontractor's insurance policy. Again, this is normal procedure. It should not cause a problem with the subcontractor's insurance carrier or agent.

Finding Subcontractors

Advertising is generally the most accepted method for attracting subcontractors. A few lines in the "classified" section of the local paper will certainly

> "The subcontractor should **carry comprehensive general liability insurance.**"

get some inquiries. The ad can be as simple as *"Snowplowers, with vehicle, needed. Routes throughout (your market) area. Call: 555-1234 for information."*

Those individuals who respond will have some specific questions and you will need to plan for how you will address them. Questions may include:

- How will I be paid?
- When do you pay?
- How do I get told when to plow?
- Where will I be plowing?

It is a good idea to put together an information packet before you place the newspaper ad or solicit subcontractors. The packet can be mailed to the potential subcontractor and will go a long way toward eliminating any misconceptions about what you are looking for and expect. The packet should include the following information:

- Your company's policy on insurance;
- A formal subcontractor agreement that explains – in detail - the subcontractor's responsibilities as well as your company's;
- The pay rate/pay scale; and
- An explanation of your company's policy on how you work with subcontractors.

Subcontractor Referral. Another method to attract subcontractors is to establish a referral program that encourages existing subcontractors to refer potential subcontractors. Subcontractors talk to one another and they get to know who the good contractors are in the market. If you pay a referral fee to existing subcontractors, and if you have treated them fairly, meaning you paid them on time and in full, the

fee can serve as a good incentive.

For example, you could pay $200 to the existing subcontractor who refers another subcontractor to come to work for your company. The condition, though, is the new subcontractor would have to work the entire season. The subcontractor would receive half the referral fee at the beginning of the winter season and the balance at the end assuming their referral worked the entire season.

Managing Subcontractors

At times, using subcontractors can be a source of irritation, so you need to realize the associated pitfalls. You also need to be prepared that a percentage of the available subcontractors will not come out when called. The reasons for these no-shows can include sickness, hangover, broken truck, no baby sitter and even the phone being turned off – inadvertently, of course.

However, subcontractors normally take very good care of their equipment since their livelihood depends on it. They don't normally ram curbs with plow trucks and then say, "Oops." And when a subcontractor's equipment breaks down, they normally work very hard to get back on the road as soon as possible rather than call and say, "My truck's broken." Subcontractors will often carry spare parts and tools to make repairs immediately so they can get back to work earning money.

When subcontractors are done for the night – or daytime, as the case may be – they should report hours (or work completed) the same day. If there are any discrepancies in recorded hours, then they can be immediately addressed. If not, you can end up arguing later on when no one remembers what went on during that particular snow or ice event. Sometimes, this means calling subs at home and waking them up even as they are trying to get some sleep. It is better to address issues right away rather than waiting.

Show Me The Money. The manner in which a subcontractor is paid is a big issue. It should go without saying that subcontractors are the lifeblood of your business – you need them to survive. Pay them on time and in full, every time. Stretching out a subcontractor only leads to a bad reputation. Some companies have subcontractors waiting in line to come work for them, simply because they pay on time and in full. This is a very big deal to subcontractors.

Compensate subcontractors for the equipment they have available for your use. Consider a "differential system" depending upon what type of equipment the subcontractor owns. Start with a base rate that is fair for your market. Then pay more money for a larger or more-efficient plow, such as the "V" blade, snow wings or "wide-out" blade. Pay even more money for high-capacity equipment, including push plow, front-end loader or dump truck.

Pay a bit more if the subcontractor carries an iPhone. These and other mobile devices can be integrated into reporting/tracking software programs that are available to snow contractors. You should then call or text them at least once during each snow event so you know they actually have the phone with them. Pay the subcontractor more money for their second and third year with your company. Lastly, factor in compensation for having a more-efficient automatic-transmission truck instead of a manual-transmission vehicle.

Subcontractor Agreements

A written contract ensures all parties know what is expected of them. Additionally, such a contractual arrangement satisfies one of the IRS "tests" for determining if a person is a subcontractor or an employee.

This contract can be as detailed as necessary. It should specifically state the subcontractor is not an employee as this satisfies the "intent of the parties" portion of the IRS test procedure and policy.

It should also detail what equipment the subcontractor will use while plowing the sites that will eventually be assigned to him. This detail should include the type of equipment, type of plow, rates charged for specific types of equipment and the subcontractor's company name.

The contractor should detail what is expected of the subcontractor in terms of availability, response time, hours available for work, any guarantees on how much work there might be available, payment terms, what documentation is necessary to be paid, the fact that no reimbursement for expenses incurred will be forthcoming, insurance requirements, responsibilities with regard to customers' expectations and other details related to the work. Remember, err on the side of caution and provide as much detail as possible.

The use of subcontractors is a tool to help your company grow. And like any valuable tool, it needs to be used effectively and maintained properly. Subcontracting is not the only way to grow your business, but it may be one of the easiest methods not requiring substantial capital investment on your part. Treat subcontractors well and they will be loyal and productive members of your team.

Chapter Eleven
SNOW MANAGEMENT EQUIPMENT

Chapter Highlights

- Finding A Plow Truck
- Selecting A Plow
- Using Heavy Equipment
- Throwers, Spreaders and Melters
- Equipment Maintenance
- Replacing Equipment

Selecting, maintaining and operating snow and ice management equipment is the most important step a snow professional undertakes. Without solid employees and reliable equipment, no customers can be served. Equipment for snow management varies greatly, ranging from the tried-and-true snowplow mounted on a pickup truck to the use of heavy equipment, such as front-end loaders and large dump trucks. Additionally, advancements in equipment over the past decade have been more significant with the addition of software technology.

Finding A Plow Truck

When it comes to putting plows on trucks, there are as many different opinions as there are equipment dealers. It can be confusing. Most large contractors will agree – for the most part – that the equipment on the market will do the job well regardless of its manufacturer.

Marketplace confusion can be traced to the addition of the driver's-side airbag in trucks. Federal law states all trucks weighing less than 8,600 lbs. gross vehicle weight must meet the same safety standards as cars. This makes automotive engineers very

conservative. Also, more than 90% of pickups and sport utilities sold today are sold to car buyers who want a "new-car ride" in their trucks, as well as all the options available in cars. Because of this, the stripped-down work truck is becoming a thing of the past.

Trucks are being designed with a softer ride and are now made closer to the ground so people can get in and out of them easier. Truck manufacturers now consider sport utility vehicles, 1/2-ton trucks and other trucks to be personal-use vehicles, not commercial-truck vehicles.

Most of the major truck manufacturers offer snowplowing packages for their vehicles. At the time of the printing of the second edition of this book, Chevrolet, Dodge, Ford, GMC and Toyota all featured trucks that can handle plows. The specifications and capabilities vary by manufacturer, so do your research before buying. Talk with other area contractors about their experiences with various models.

For complete information on what truck manufacturers have to say about mounting plows and available models, visit your local dealer or visit the following web sites:

- www.chevrolet.com

SNOW FACT:

WINTER STORM OUTLOOK
Is issued prior to a Winter Storm Watch. The outlook is given when forecasters believe winter-storm conditions are possible and is usually issued 48 to 60 hours in advance of a winter storm.

- www.dodge.com
- www.ford.com
- www.gmc.com
- www.toyota.com

Heavy Duty. As you can see, truck manufacturers are saying that if you want a plow, get a 3/4-ton truck or larger. From a practical point of view, this is good advice. The primary concern is the amount of weight placed on the truck's front end. For example, if a pickup is designed to carry three passengers, that weight must be figured into the equation to see if the front axle is overloaded. Overloading should not be taken lightly since it can affect break wear and federal safety standards.

But what if you plow alone and are not carrying the weight of additional passengers? This is not considered in the equations. Another point not being considered is rear ballast. The more weight – within limits – added to the rear axle subtracts weight added to the front axle by adding a plow.

The question then arises, "What can happen if you put a plow on an unapproved truck?" The manufacturer's warranty on the front-end suspension will be voided and the brakes may also prematurely wear, causing safety and liability concerns.

There are dealers who will install plows on almost any vehicle out there, but I don't recommend doing so. Using a plow on an unapproved truck creates serious liability and safety concerns in the event of an accident.

2WD vs. 4WD. Another concern is whether a plowing vehicle needs to be four-wheel drive (4WD) or two-wheel drive (2WD). Many states departments of transportation (DOT) use 2WD trucks, often with V-box salt/sanders mounted in the beds. For DOT work, the 2WD truck is fine. The vehicle is always moving and inertia assists in keeping it from losing traction.

Private snow management contractors work primarily in parking lots and driveways, which require frequent stops, starts and changes in direction. In these instances, it is almost always beneficial to use 4WD vehicles instead of 2WD vehicles. Additional maintenance may be involved given the nature of having 4WD hubs, a front drive shaft and from the inherent problems that arise from plowing snow.

4WD vehicles move through deep

TOP SNOWSTORMS IN UNITED STATES' HISTORY

#4 BLIZZARD OF 1978

Jan. 25-27, 1978

A tremendous blizzard struck Ohio, Michigan, Kentucky, Illinois, western Pennsylvania and southeast Wisconsin. One to 3 feet of snow was common throughout this area with 50- to 70-mph winds whipping up 10- to 15-foot drifts. Milwaukee measured nearly a foot of snow from this large and very intense storm system. Ohio was hardest hit with 100-mph winds and 25-foot drifts. Much of the affected area was paralyzed for several days. Overall, more than 70 deaths were blamed on this storm.

Source: National Weather Service (www.crh.noaa.gov)

Selecting the proper truck and plow requires some research into the various models and options.

snow with little additional effort and will allow the stacking of snow much easier than with a 2WD vehicle.

Electrical Systems. Electrical systems on plow trucks need to be carefully evaluated, as snowplows draw considerably more power than standard truck equipment. Some plow manufacturers recommend alternator units that put out a minimum of 74 amps. Even this may not be enough in some situations when you add lights, a safety bubble or other plow accessories. Many plowing contractors add a second battery and upgrade the alternator significantly when adding a plow to the vehicle.

Transmissions. Transmission selection can lead to spirited discussions among snow professionals. While some prefer a standard transmission, most will agree that an automatic transmission is more efficient during plowing operations. Automatic transmissions allow for faster reverse speeds. Most automatic transmissions will also not require you to come to a complete stop prior to shifting into

reverse. Standard transmissions will grind incessantly if you attempt to shift into reverse prior to stopping the vehicle.

Plowing with the automatic transmission in "low" is recommended because if the shifter is in "drive" or "3," the transmission will automatically try to shift up or down depending upon RPM and load. It is possible the transmission will go through all the gears - sometimes several times – during one plow run down a parking lot depending upon the weight of the snow and the resistance it creates.

This constant shifting is hard on the transmission and can greatly shorten its life. Plowing only in "low" will not allow this constant shifting and pounding of the linkage. This is not to be confused with the "low" range on the 4WD shifter. "Low lock" on the 4WD shifter is for use in very, very deep snows. Most of the plowing done will be in "4WD high" with the transmission shifter in "low."

If the truck is outfitted with "Overdrive," the overdrive must be disconnected prior to plowing. With the overdrive engaged, plowing will quickly burn out the transmission. This is an expensive lesson that is avoided by paying attention when starting up the vehicle.

Even if you are careful, the automatic transmission will heat up during plowing operations. If the truck is not equipped with one, installing a transmission cooler is a wise investment. Keeping the transmission cool should be a top priority when plowing with a truck.

Leafs vs. Springs. According to Chuck Smith, owner of the Internet site www. snowplowing-contractors.com and considered one of the foremost authorities on plow trucks and related equipment, mounting a plow on a vehicle with torsion bars is not recommended.

He also recommends avoiding putting plows on vehicles with coil springs in the front end. Leaf springs are much better, as they are designed to carry heavy loads. Leaf springs can be easily replaced and repaired because nearly every spring shop can make new ones or replace broken leaves. You can also add a leaf to compensate for the dip the front end takes with the additional weight of an attached plow. If forced to put a plow on a vehicle with coil springs, consider adding an airbag to the inside of the coil to alleviate the forced compression that accompanies placement of a plow on the truck.

However, Smith reminds contractors that adding leaves does not increase the front axle's carrying capacity. The extra leaf compensates but doesn't strengthen the axle in any way. The axle can only

A variety of snowplow types and features are available, including (clockwise from left to right): V-plow, specialized mounting systems, polyurethane edges, rear-mounted "pull" plows and trip-edges.

carry what it was designed to carry.

Selecting A Plow

There are some important decisions to make when purchase a plow. The amount of money you have is a significant factor in the decision-making process and like most things you get what you pay for.

Plows are manufactured for specific uses. Light-duty plows are inexpensive, and while usually adequate for residential plowing situations, they often don't perform well in commercial plowing operations.

Straight blades are less expensive than V-blades, but they have limitations on what they can do efficiently. Straight blades must be angled to be productive, Pushing "straight on" with a straight-blade plow is the least-efficient method of moving snow. When you add a set of snow wings to the conventional straight blade, you will increase efficiency by 35% but you are adding weight to the front end of the truck.

Purchasing a V-blade plow can cost as much as an additional $1,000, but it is usually twice as efficient and well worth the investment. V-blade plows take more skill to operate, and it can take as long as a season to get used to the controls. Again, the increased efficiency makes the slight inconvenience worth it.

Expandable plows have become prevalent in the industry, and advancements in the technology have increased productivity considerably. Almost all plow manufactures offer a model that has hydraulically operated wings that widen the plow by 2 to 3 feet. Additionally, these automatic wings tilt forward slightly to allow for "scooping" snow as the plow moves forward. The overall cost has become more affordable as the technology has advanced.

Let's Make A Deal. The most important thing to consider when selecting a plow is the dealer support you receive following the purchase. Some dealers will promise the world and then will not have time for you when there is a problem. Buying a cheap plow can become quite expensive if you cannot get replacement parts or if the dealer is only open a few hours during a snow storm. Ask the dealer for references and talk with other snow professionals about their experiences. Having a reliable dealer to work with can offset the expense of the initial purchase.

Don't Trip Up On Trip Blades. A "trip" feature on a snowplow blade allows the plow to give when passing over raised objects, such as manhole covers. The feature is typically available as full-trip or bottom-trip blades. Ask a contractor who has been using bottom-trip blades his entire career and you'll find that the bottom-trip plow is the best. Ask the same question of a contractor who has used nothing but full-trip blades and you'll usually get the same answer. One is not better than the other. Instead, it comes down to personal preference.

Putting A Push On Snow. In the late 1990s, a company in Rochester, N.Y., patented a unit designed to push snow using loaders of various sizes. This increased efficiencies for this type of equipment five-fold. Once patented, snow pushers became more commonplace in the New York and Pennsylvania snowbelt areas. Over the past decade, snow pushers have become quite commonplace in the commercial snow industry. It is the rare snow contractor

Plow selection can impact your crew's efficiency and your bottom line.

who still uses loaders, backhoes or skid steers without having a pusher on the front of the unit.

Many contractors have been drawn to these box plows because they can remove more snow per pass than traditional plows. This increases efficiency, decreases plowing time and allows the contractor to serve more accounts. This means more revenue in addition to more profit.

Several brands of box plows exist in the market and they come in a variety of sizes and quality levels. The best box plows are constructed of heavy-duty steel with a properly engineered support structure. Such units can move hundreds of cubic yards of snow in a single pass. In deeper snows, their efficiency increases more dramatically.

Box plows should be used on loaders *only*. At one point, some contractors argued that box plows could be utilized on trucks. This proved to be a bad idea. The box plow is, by nature, constructed very solidly and must be mounted directly to the frame of a truck. Because of the solid construction, any energy shock from a front-end collision – even with a man-

hole cover – is transferred directly to the frame of the truck. In some instances, there can be a failure of the structure, most likely the frame of the truck. This is an expensive repair. As such, these units quickly fell out of favor with professional snowplowing contractors.

As mentioned previously, box plows are for use on loaders only. They come in all sizes, from 6-foot models for use on skid steers to 30-foot-wide models that are used primarily for airport applications. Box plows are easy to mount on a loader and increase efficiency greatly. In per-push or seasonal pricing, they can increase margins dramatically and are becoming the equipment of choice on large properties.

The sectional box plow is a variation of the original concept and a decided improvement on the traditional box plow. The sectional box plow was developed by Chicago snow contractor Randy Strait. The sectional box plow was patented and made available to the general snow contracting business. Those using them swear by its ability to actually scrape snow from the pavement surface. A traditional box plow cannot do this due to its solid moldboard. The sectional box plow is essentially cut into pieces so that "sections" of the mold-board move up and over obstructions on surfaces. Anchored curbs are left in place as each section raises up and floats over the obstruction. The steel cutting edge, generally not feasible to use on tra-ditional fixed edge pushers, moves out of the way of the obstruction, eliminat-ing the impact/force transfer to the rest of the unit.

Structural integrity is not compro-mised in these situations as it is with traditional box pushers. Uneven pave-ment is cleared more efficiently since the moldboard "molds" itself to the configuration of the pavement surface. However, such improvements come at a cost. These units are significantly more expensive to purchase, but less expensive to maintain. The result is a much cleaner pavement surface, especially on an un-even surface.

Poly Edges. Another recent advance-ment has been the use of polyurethane

wearing. Polyurethane edges don't come apart and are hard and thick enough to retain some cutting ability. Due to the sliding action, a polyurethane edge will not disturb gravel parking surfaces near-ly as much as steel. Redistributing gravel after the snow melts can be an area of contention between property owners and contractors. Using a polyurethane edge virtually eliminates this problem.

Probably the most important rea-son for using a polyurethane edge is that it is very forgiving. Parking lot surface imperfections will not be disturbed by polyurethane, while steel may peel up the pavement the imperfection. Poly-urethane does not disturb manhole cov-ers, cracks in the pavement, grates that stick up slightly and pieces of cracked concrete.

Polyurethane will "remember" its original configuration and return to its straight edge after bending to go over surface imperfections. Eventually, poly-urethane will be the cutting edge of choice and steel will be the cost-cutting option.

"Probably the **most important reason** for using a polyurethane edge is that it is very forgiving."

cutting edges on plows and pushers. Numerous plowing contractors are re-placing steel cutting edges with polyure-thane when the steel wears out.

The reason is polyurethane cutting edges save snow professionals money. While the initial expense of the poly-urethane edge is several times more than that of steel edges, polyurethane lasts as much as five-times longer.

When you see a plow truck going down the road with sparks flying, it is the steel cutting edge coming apart or

Using Heavy Equipment

Long gone are the days where snow-plow operators relied solely on pickup trucks to remove snow. Today, many successful companies use a wide vari-ety of so-called "heavy equipment" in snow management.

The most commonly used units are skid steers, loaders, backhoes, payloaders and dump trucks. Often, this equipment is already available to a snow contractor from his summer-time business. This is true in the case of landscape, pavement maintenance and construction industry professionals. Many contractors also successfully form subcontracting agreements with companies – residential or commercial builders – who can supply heavy equipment and the appropriately trained operators.

Using or subcontracting the use of this heavy equipment can greatly boost your snow management efficiency. This equipment can be mounted with large plows or snow throwers that can move substantial amounts of snow in a relatively short amount of time. However, this equipment does require extra training and maintenance.

The proper way to use such equipment is a book in itself. Suffice it to say, using heavy equipment to remove snow on large, commercial sites can significantly boost profits. However, make sure you are well versed on the proper, safe-operation methods of such equipment, as well as the ongoing required maintenance.

Today, many companies use "heavy equipment" for snow removal. Pictured (top) is a loader with a box plow attached and (bottom) a skid steer with a straight-blade plow.

Throwers, Spreaders and Melters

There is other equipment that contractors can employ to clear snow and ice from properties.

Sweepers/Snow Throwers. There are situations, clearing sidewalks for example, when smaller, more maneuverable equipment is needed. In these situations contractors often use walk-behind snow throwers and brooms, tractor-mounted sweepers or three- or four-wheel all-terrain vehicles – ATVs or quad runners – mounted with plows. In addition, recently developed articulating 4WD tractors handling numerous attachments specifically designed for the snow industry.

This equipment allows contractors to quickly and efficiently clear snow from otherwise inaccessible surfaces. Such equipment is much faster and often much more profitable than using crews to physically remove the snow with shovels. Sidewalk snow clearing will be discussed in greater detail in Chapter 13 *(See page 135).*

Deicing Spreaders. We will discuss deicing strategies in greater detail in Chapter 14 *(See page 143),* but there are some basic points to discuss here. Deicing equipment can range from the simplest hand-powered spreader to a sophisticated, electronically controlled liquid tank and application system mounted on a commercial truck.

The most commonly used deicing equipment for snow removal professionals is the back-end-mounted hopper/spreader or the slip-in V-box spreader. A wide variety of sizes and holding capacities are available to contractors based on budget and spreading requirements. Price can also vary

Snowmelter technology has allowed portable units to become more common with contractors.

depending on the mechanism used to control the rate at which deicing material is spread. The most convenient units have electronic controls mounted inside the vehicle's cab. The hopper/spreader is mounted on the pickup truck's bumper and can hold around 9 cubic feet of material.

The V-box spreader is mounted inside the truck bed and can hold more material. Load capacity is determined by the size and weight of the vehicle in which the V-box is mounted and varies from 1 to 7 tons. In the past decade, manufacturers have developed polyurethane V-boxes to handle the weight and stress that comes from holding several tons of deicing material. This advancement has lengthened the life of such equipment.

Snowmelters. Over the last several years, advances in equipment have included one new item. In the past, when snow needed to be relocated, the only way to accomplish this was to place the snow in a dump truck and haul it

to another location. This often required traveling long distances, especially when working in a crowded downtown area, such as Chicago. Metropolitan areas are so tight for space, there is no place to put the snow and dumping into rivers and streams is no longer legal.

I became bothered by the lack of options when performing all the snow-management duties at the 2002 Winter Olympic Games in Salt Lake City. When exploring alternatives to hauling snow off-site, using snowmelters came to mind. At that time, in-ground melters were in use at airports. Likewise, portable units were environmentally unfriendly and not a viable option given the strict parameters in place for the games. It became apparent hauling was really the only viable option, no matter how expensive.

After 2002, snowmelter technology was available that catered to a contractor's specific needs. Using heat-transfer boiler technology, environmentally friendly portable melters were developed that discharged clean water and could easily be transported between job sites. No longer confined to municipal and airport applications, snowmelters are now more commonplace in the contracting business.

From an economic standpoint, if the round trip time to haul snow by truck – from starting to fill the truck to the time refilling the truck begins again – exceeds 45 minutes, then melting becomes economically viable. As cycle times for trucks increase past 45 minutes, it becomes exceedingly cheaper to melt rather than haul it away. Additionally, water discharge from melters can now legally enter local ecosystems, including streams and rivers in many parts of the country. However, check regulations in your service area before doing so.

Equipment Maintenance

Equipment-maintenance costs can easily get out of hand if you ignore preventive-maintenance issues. Develop a formalized preventive-maintenance plan for your company because being proactive will save you thousands of dollars in emergency repairs. This can be as simple as posting reminders for lube, oil and filter service on your computer calendar so reminders pop up regularly.

Oil Changes. In the winter, it is highly recommended that you do a lube, oil and filter service every month. This might seem excessive, but you are working the truck very hard when plowing snow. Treat the truck right and it will last much longer. The only exception would be in markets where only light snow accumulations occur. Those contractors might use their truck only once or twice during the winter and wouldn't need to change oil as frequently.

Transmission Fluid. A snow professional should change transmission fluid more frequently – about ever 12,000 miles – especially if they have come through a hard snow event or a series of snow events. Transmissions get hot during plowing operations and the fluid loses its viscosity. Changing the transmission fluid frequently and adding a transmission cooler will extend your transmission's life.

Hydraulic Fluid. Any leaks in the plow unit's hydraulic system is a cause for concern because water can seep in where the leak occurs. Leaky piston seals, couplers and connectors can allow water into the system. If not drained, that settles to the bottom of the reservoir causing rust to form on internal parts and surfaces. Check often and address leaks as soon as noticed. Lastly, change the hydraulic fluid at the end of the winter season so you can start with fresh fluid in the fall.

Grease. Grease is very important. Check universal joints every two snow events and grease as needed. Grease the hinge joint on V-blade plows after every plowing run. Do the same for the hinges on fold-out snow pushers and on sectional plows. All other grease fittings on plowing equipment should be checked regularly. For loaders and skid steers, grease the fittings daily. It cannot be stressed enough just how important this task is ensuring reliable equipment performance.

Salt Spreader Maintenance. Preventive maintenance on salting equipment is also very important. Salt's corrosive nature demands that care be taken with spreaders' moving parts, therefore they should be cleaned thor-

Rear-mounted spreaders are commonly used by contractors to apply de-icing materials.

oughly after each use.

After cleaning, apply a fine film of an anti-corrosive mixture. This will go a long way toward keeping moving parts in proper working order. To protect parts from corrosion, some contractors use a mixture of diesel fuel, water and dishwashing soap (60/30/10) sprayed from a one-gallon pump sprayer on all surfaces that come in contact with salt. Only a few ounces of liquid material are required to adequately cover all surfaces, and this allows the salt to flow smoothly out of the hopper.

Your equipment represents a significant investment both at the time of purchase and when you must make repairs. Most contractors know and understand that preventive maintenance costs much less than having to make major repairs.

Replacing Equipment

Contractors struggle with the question of how long to keep equipment and trucks in service before replacing them. There is no hard-and-fast answer to this question.

A well-maintained truck, even with significant plowing hours, should last at least five years. However, 10 years is easily obtained with a proper preventive-maintenance program in place.

Track all operational and maintenance costs for the vehicle in question. If the combined costs in a given year exceed the payments and operational costs for a new vehicle, then it's time to consider replacement.

Plows are a different story. It is generally accepted that a plow's operational lifetime is around six or seven years, maximum. Box plows can have a virtually unlimited life if maintained and not damaged by careless loader operators. Salt spreaders will last eight to 10 years – excluding units using engines. Hydraulically operated spreaders will last considerably longer if maintained.

Chapter Twelve
PLOWING EFFECTIVELY

Chapter Highlights

- Selecting The Right Equipment
- Efficient Snowplowing Techniques
- Plowing Safety
- Avoiding Costly Mistakes
- Employee Training

Contractors who have been plowing snow for several years have developed systems for plowing that are quite adequate. It would be presumptuous to think I could improve upon these established methods for moving snow. However, there are some suggestions that might assist novice plowers to better manage the snow on the sites they service.

Those who plow snow on a per-hour, per-truck basis have little incentive to move snow on customer sites efficiently and effectively. Experience shows that those who plow snow on a per-push, per-inch, per-event and per-season basis inherently seek out the most efficient way to clear snow. This is done to move from site to site as quickly as possible to increase production times and profits. For this discussion, let's assume the plowing contractor wishes to increase production times to increase revenues and profits.

Selecting The Right Equipment

In Chapter 11 *(See page 107)*, we discussed the importance of selecting the correct equipment to match the snow management job at hand. The proper use of that equipment plays an important role in the efficiency and profitability of your work. I encourage you to think outside the box when detailing how your crews "attack" a site during a snow and ice storm.

Part of the problem is that, once a mind-set is established, it becomes quite difficult to change and go to the next level. Always consider the more efficient ways to clear snow from an account's property.

For example, most contractors who use snow pushers are amazed at the increased productivity they offers. However, these same contractors will continue to plow smaller lots with pickup-mounted plows instead of using pushers to increase productivity.

Another example is suggesting to a contractor that a 20-foot-wide, fold-up snow pusher can be used in a typical fast-food restaurant. A closed-minded person will believe this is a crazy idea, but in reality it is being done in several markets. Imagine plowing a fast-food restaurant in 10 minutes using a 20-foot pusher and a plow truck. Now imagine the profit that results if you are charging per-push.

The plow truck pulls snow out of the tight corners into the main lot, and then leaves the site. The loader and 20-foot pusher come in and make two passes around the building, totally clearing the lot. Of course, this will

Heavier equipment, such as skid steers, are often used by contractors to remove snow from accounts.

not work if you must drive the loader and 20-foot pusher down the road to get to the site. However, if this same fast-food establishment is located in an out-parcel immediately adjacent to a mall complex, this configuration works nicely.

The use of a 10-foot pusher on a 4WD backhoe will also do the same fast-food restaurant all by itself – assuming an open-minded and capable operator is at the helm. Pushers on 4WD backhoe loaders will back drag, get close to curbs, tuck corners and pile snow just like plow trucks, only much faster.

Efficient Snowplowing Techniques

Let's assume you still do not use snow pushers and instead use plows on trucks. In parking lots where the lot completely surrounds the building, the old-fashioned method would dictate plowing each side of the building in turn. When diagramed, one can ascer-

tain that this is an effective method of plowing.

You move the snow from the lot on one side of the building and then clear the second side by backing up onto the already cleared corner of the lot. This is repeated around the building until the entire lot is cleared. Unfortunately, this method results in considerable backing up of the plow truck – an inefficient use of time. Going forward as much as possible, with the plow down and pushing snow off the pavement, is preferable. However, old-school thinking won't allow for other ideas to come into play.

How else can you clear the lot if you don't push the snow to the outside edges one side at a time? Envision, if you can, the lot is clear of vehicles and that the interior nooks and crannies are "pulled out" into the lot past any islands or other close-in obstructions prior to doing any forward plowing.

Once the back dragging is completed around the building, you can begin running around the building in circles – counterclockwise in this example – angling the blade to the passenger side of the truck.

With each concentric circle around the building, you go in wider and wider circles, pushing the snow closer and closer to the perimeter of the parking lot. Essentially, you are "wrapping the building" to clear the snow further and further away from the building.

Eventually, you may have to go around light poles in the lot, but by cutting away from the base of the pole, you'll eventually get out past that pole and continue your concentric circles, further advancing the snow toward the perimeter of the lot. Once the majority of the snow is at the exterior curb lines, you can push in the entrances to the lot (from the street), and tuck the corners to neatly finish off the plowing.

Experienced plowers will adamantly say a contractor can cut the plowing time by nearly 50% in this fashion. Exact time reductions will depend on the number of obstructions in the lot, as well as snowfall restrictions. However, if the above example is followed closely, dramatically reduced plowing times and increased production will result.

Explanations Of Plowing Patterns. There are several tried-and-true plowing patterns that contractors may consider using in accounts. They are describe – with diagrams – in this chapter.

When Arriving On Site. Some examples of things contractors need to be aware of when entering a lot to start plowing operations – in no particular order – include:

- Ascertain where catch basins are located so you don't inadvertent-

ly create a dam that might cause flooding.

- Be aware of curb locations so you don't damage your plow (or the curb) during plowing operations.
- Know where fire hydrants are located and never plow snow onto a hydrant. Firefighters may know where the hydrant is located anyway, but making them dig out the hydrant in an emergency slows down their response time.
- Don't plow snow onto sidewalks intentionally. Most walks are to be kept clear for pedestrian traffic.
- Know if the customer allows plowing snow onto shrub beds or planted islands.
- Watch for imperfections in the parking lot that can cause head banging, so-called "wake-up calls," if struck while plowing. This not only creates severe headaches, but can cause injuries that require stitches. A site visit prior to the onset of the winter season will allow you to identify where these imperfections are located.
- With automatic transmission vehicles, always plow in "low" on the column, even if the run is short. This does not mean plowing in 4WD low. Do not plow in "drive" as the transmission will attempt to shift gears as the resistance to the truck increases or decreases during plowing. You can quickly burnout the transmission or transfer case on your truck by not plowing in low.
- Always disconnect overdrive switches and never plow with overdrive engaged. You can burn out the transmission in minutes.

One Cautionary Note. Always plow

SNOW FACT:
SNOW SQUALL
A brief, but intense, fall of snow that greatly reduces visibility.
It is often accompanied by strong winds.

121

SNOW PLOW METHODS

THE TANK TURN

A more productive variation of the old go-ahead-back-up routine, the tank-turn method places the pusher on the pavement 90% of the time. This pattern should be used when snow can be placed at either end of the plow run, or when there are obstacles to work around in the parking lot. It can also be used when the lot is too big or the snow is too deep for the Zamboni method (page 128) to be effective.

Push the snow to one end of the parking lot and into the pile. Remember, don't stack the snow higher than the top of a pickup truck cab. Then back away from the pile, do a three-point turn, move over approximately a pusher width and then head back the way you came with the pusher down. Production is improved because you are not backing up to where you started, and backing up with the pusher in the air is not productive.

Caution: Snow pushers are just that – pushers. If they were designed to stack snow, they would be called "snow stackers." If snow is to be stacked, bring in a loader with a bucket to stack the snow, and then charge the customer accordingly.

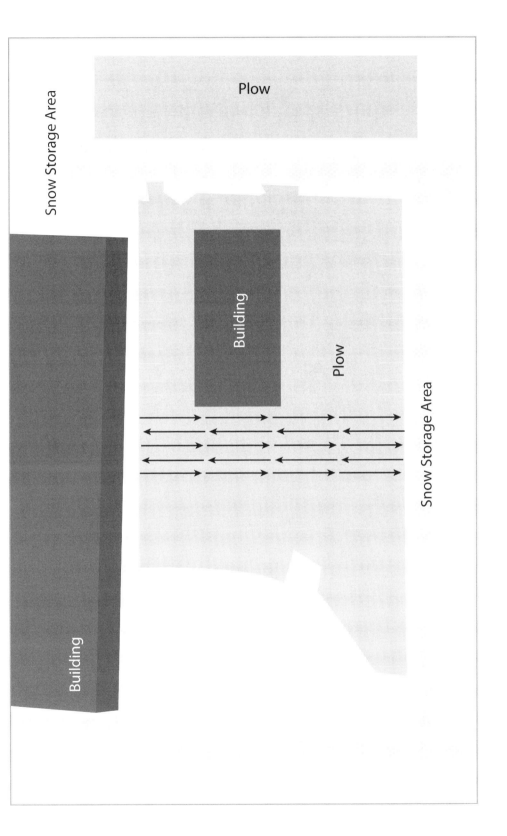

SNOW PLOW METHODS

THE U PATTERN

Assuming the site is set up properly, and you can only deposit snow at one end of the lot, start your plowing pattern where the snow is to be deposited. In most cases, this is a point far away from the building. Head toward the building with the pusher on the pavement and make the turn at the building. Keep the pusher down throughout and run about halfway along the building and – with the pusher still down – turn toward the area where snow is to be piled (pusher still down).

Make the run to the pile area, the pusher is still down and filling up quickly. Deposit the snow at the authorized pile area. Then, go back to the starting point, moving over not quite one-pusher width (toward the center) and begin the process again, moving not quite one pusher width at the center turning point. You will leapfrog the cleared area and clear two halves of the parking lot at once, all the while keeping the pusher on the pavement 75% to 80% of the time.

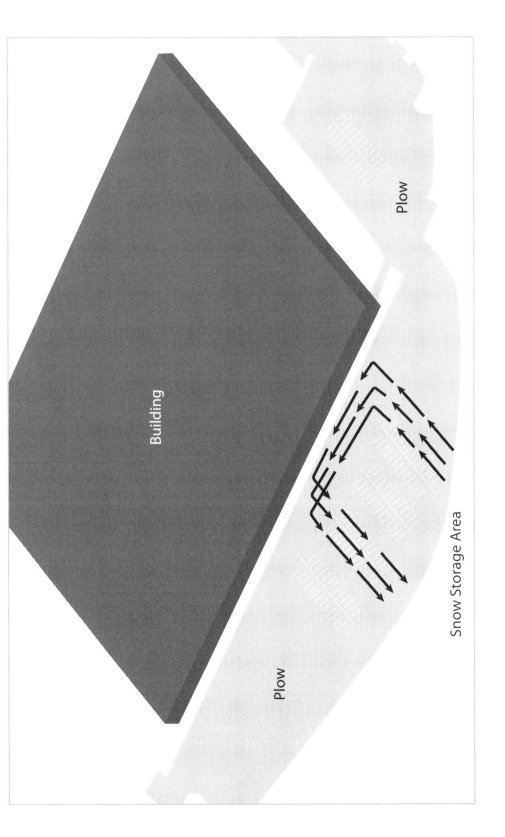

Plow

Building

Snow Storage Area

Plow

SNOW PLOW METHODS

THE WRAP

This method is done when the parking lot or driveway surrounds the building. The plow truck enters the site, plows out all the interior areas, loading docks, wide entrances, small parking areas into the main-drive area. Next, the plow truck runs around the building with the blade angled away from the building. The truck "sweeps" in and out to avoid islands or other obstructions that stick out. Then, after going out as far as it can in concentric circles, the truck goes back and "tucks the corners" and the end of a stream of parking areas. Finally, the entrances to the site are plowed and cleared.

A three-quarter wrap is done in the same fashion as the full-wrap method but when there are parking and drive lanes on only three sides of the building. This requires the truck to perform a tank-turn method (page 122) or three-point turn at the end of the run, and then sweep back in the same direction it came from, moving further away from the building with each pass.

Sometimes you can wrap the building when only two sides have pavement. This would be for large sites only, as small sites typically don't allow for traveling around the building with the blade on the pavement.

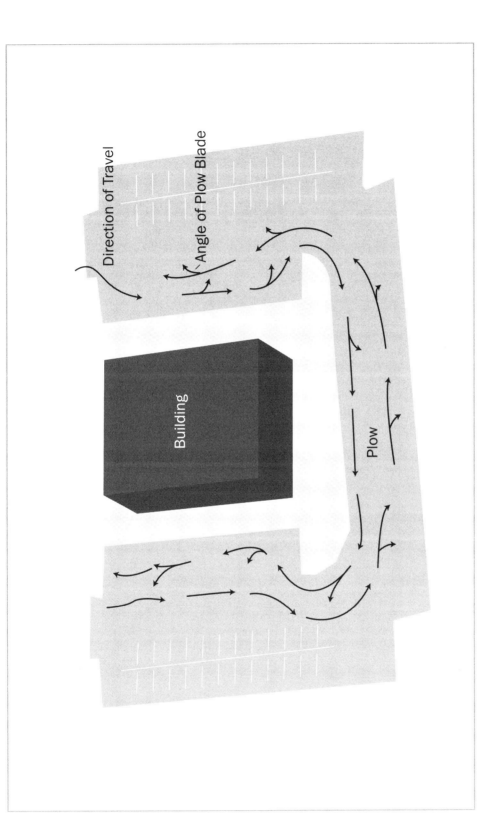

SNOW PLOW METHODS

THE ZAMBONI

The Zamboni method is done when the parking lot is elongated and allows for dumping snow at each end. Just like a Zamboni clears an ice rink, the loader/pusher combination runs in elongated circles, moving over one-pusher width with each pass. At either end of the lot (each "goal end"), the pusher is lifted up while the loader continues moving forward, essentially dumping the snow during the turn at the end.

This results in quite a mess and requires one or two passes along each end of the parking lot to get it cleaned up. The idea is to keep the loader moving forward at all times with the pusher remaining on the pavement 80% of the time.

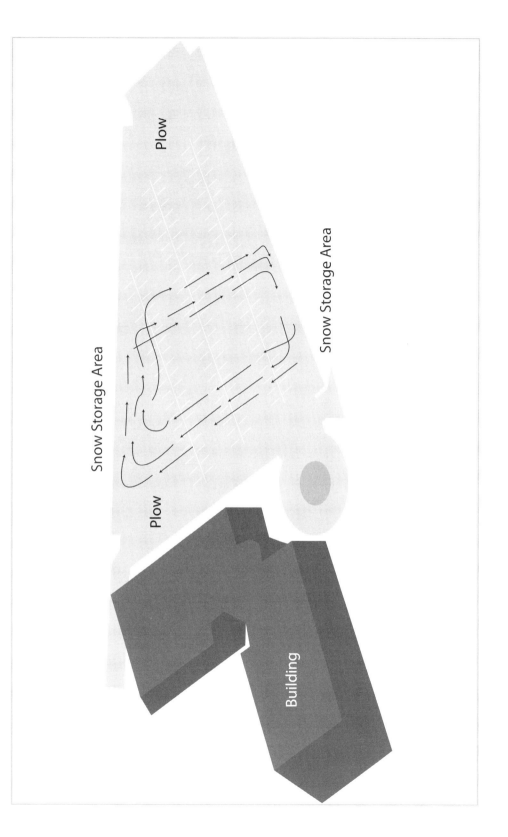

For large commercial accounts with large parking lots, heavy equipment may need to be used.

with the seatbelt securely fastened. It is inevitable that you will hit an obstruction at some point in time. It may be a piece of pavement sticking up or a manhole cover that is up a half inch. The truck may come to a dead stop in such instances, but the physics of inertia will allow your body to continue its directional path. It usually takes one time of "kissing" the steering wheel to reinforce this edict.

Tips For A Quality Job. Some important tips contractors can use for providing a quality job for the customer include:

- Clean up after yourself in the lot. Don't leave "feathers" or "trailers." A clean lot is what most customers expect even if you have not specified this in the contract. Lines of snow from feathers make the lot look messy, and give the impression the plower doesn't care about the quality of the job.

- In lots that don't completely surround the building, try to plow in straight lines. It makes for a much neater appearance and shows a level of professionalism.
- Cut corners tight and square them off. Veering off at the end of the run without squaring off the corners gives the appearance of not caring about the finished product.
- At the entrances to the lot, refrain from piling up snow at the corners of the apron. This can restrict vision when cars attempt to exit the lot and you could be held liable for creating a dangerous situation if an accident occurs.
- Don't intentionally "plow in" cars that may be parked in the lot. Even if the car shouldn't be there, you may be considered liable for damages.
- Keep parking lot aprons free of excess snow so vehicles may easily enter and exit from the lot safely.

Plowing Safety

Snowplow operators are asked to work in very hazardous conditions, often late at night and in difficult circumstances. As a result, injuries, vehicle accidents and damages to the customer's property can occur if safety is not a top priority. Also, being unprepared for equipment breakdowns can create dangerous situations in which the plow operator is exposed to the elements or vehicle traffic while trying to fix the broken machinery. Some important safety and preparedness tips when plowing or getting ready to plow include:

- Check all fluids regularly, including the plow fluid, engine oil, transmission fluid and washer fluid.
- Check tire pressures regularly – overly hard or overly soft tires can hamper efficiencies.
- Carry spare pins, hydraulic fluid and a spare hydraulic line in case one blows during plowing operations.
- Carry tools necessary to change a hydraulic line, add fluids, or to conduct minor repairs out in the field.
- Carry spare clothes just in case you have to get down under the vehicle while it's snowing and you get wet. Getting sick is not an efficient way to continue generating income while it's snowing.
- Carry a roll of toilet paper. Trust me on this one; you'll be very happy you have it.
- All plowers are convinced that they will never get stuck. However, in the unlikely event it happens to you or some other poor sap, carry a tow chain, or better yet, a cloth "snatch strap."

- Have gloves with you – you'll be very glad you did if you ever get stuck during a storm.
- A few extra bungee cords can come in handy. Place a couple under the seat – just in case.
- Fluids, such as water or a sports drink, can be important during long stretches in the truck. Obviously, beer or any alcohol is never appropriate while operating a vehicle or piece of equipment.

Avoiding Costly Mistakes

The following are some common mistakes plowers make when first starting out. Unfortunately, some of the most experienced plowers can also make some of these mistakes.

- Taking too big a bite when plowing – this allows for feathers to develop off the leading edge of the plow.
- Trying to stack the snow too high with the plow. This can damage the plow frame and can also cause the truck's front axle to get hung up in the snow pile, requiring the use of that chain you didn't think you would ever need.
- Not asking for help when or if you do get stuck. It is amazing just how much camaraderie there is among contractors when they are out there fighting the storm. Assistance when you are stuck is usually just a short distance away.
- Not taking a moment to ascertain the most efficient plowing pattern for the lot you are plowing. Just digging in and getting it done can lead to double plowing and inefficiencies in your plowing pattern.
- Being in too much of a hurry – customers become unhappy when you do a hurry-up job. You are

SNOW FACT:

The old saying that 10 inches of fresh snow contains 1 inch of water is only occasionally true. In reality, 10 inches of snow can contain as little as 0.10 inches of water to as much as 4 inches.

a professional and should treat every job and every site as if it is your most important project. Remember, you're never too busy to do the job right.

- The job is never done until the paperwork is completed. Keep track of what you do and write it down. After a long stretch in the truck, it is often difficult to remember what was done in the heat of battling the snow and ice storm, but it is important.

- Not documenting conditions while you are plowing. Inevitably there will be insurance issues that arise down the road and try-

Landscape contractors often have a natural avenue into the snow business.

ing to reconstruct a snow event 18-months later is difficult.

- Don't panic when it is snowing hard. If necessary, take a deep breath, review the situation, make a plan and follow it.

Employee Training

Experience is the best educator when it comes to plowing snow safely and ef-

fectively. Being efficient and methodical can make life much easier during a snowstorm. Unfortunately, as your company grows you will be less and less involved in the actual plowing of accounts. So you will have to learn to rely on your employees to get the job done. This means you will need to conduct adequate training so employees serve your customers with the same quality level you would if you were out there plowing.

Thorough instruction and training are key ingredients to successful and safe snowplowing. For example, instructing loader operators that their buckets need to be in the float position and not to attempt to apply down pressure to the rear of the pusher is key to successfully operating these units. The tendency for any loader operator is to apply more down pressure to cut into the snow. However, these units are designed to slide over the surface to adequately remove the snow in front of the unit. In the case of sectional pushers, applying down pressure can snap, or damage, one or more of the lower rubber spacers. This can render the unit incapable of continued efficient operation.

Plow truck operators also require training to adequately and safely do their appointed jobs. Never assume the new driver instinctively knows how to plow snow. Start by showing them a training video that details the proper pre-storm preparation, proper methods of plowing smaller properties and what to watch for when plowing larger sites. There are numerous training videos on the market that snow contractors can use as part of a formalized training program. Check with industry trade associations, magazines, manufacturers or distributors to learn what's available.

HOW TO SET UP AN EFFECTIVE IN-HOUSE TRAINING PROGRAM

Setting up an in-house training program for your employees is not difficult, if you know what needs to be accomplished. Here are some suggestions snow and ice contractors can use to set up a training program for their employees:

- Do an outline of what is to be covered and be consistent in your message.

- Have separate meetings for employees and subcontractors. You do not want to "mix" the two for fear of having the subcontractors classified as employees. The tax and workers' compensation implications can become prohibitively expensive.

- Go over the plow and controls on a real truck. Let each employee and subcontractor operate the plow and demonstrate the different positions to be used when in operation.

- Go over in-the-field repairs that might be necessary. Remember, new plow drivers know nothing. They have no idea what might happen. They will just bring the truck back to the shop and say, "It's broken." Minor repairs can be made in the field, and employees need to be made aware of what to do in these situations.

- Explain on-site methodologies and how to "look" at job sites when determining best practices when plowing.

- Explain about "wake-up calls" – manhole covers sticking up, broken concrete, obstructions – that can damage plows and cause personal injury from sudden stops.

- Do your training every season with newbies and veterans. Getting both together allows for interaction and camaraderie – you can learn from peers.

- Document when your training took place and what was covered. This is evidence of a formalized training meeting. In the unfortunate circumstance of a slip-and-fall incident, you may need to prove you are doing your best to keep your employees and subcontractors informed and educated.

On The Web

Do you need to order another copy of "Managing Snow & Ice," or could you use a CD of snow management business forms to organize your financial data, improve your contracts, subcontractor agreements and sales proposals? For access to these and additional resources on snow and ice removal, visit www.snowmagazineonline.com

Chapter Thirteen
SIDEWALK SNOW MANAGEMENT

Chapter Highlights

- Crew Management Strategies
- Small Crew vs. Large Crew
- Sidewalk Snow Management Equipment

As your business grows, another service area you can consider expanding into is sidewalk snow management. If your company doesn't already offer this service, it is likely a customer will eventually ask you to provide it. However, you need to pay close attention to details and productivity goals for this to be a profitable portion of your business.

In most companies, sidewalk snow management demands a significant invest of time and hands-on management. However, it can generate higher margins, in some cases, than snowplowing. It takes strong organization and goal-setting skills to achieve maximum performance from your crew(s) and hit your profit goals.

Managing performance in snow and ice management organizations – whether on-site or at multiple locations – is a complex issue. With sidewalk snow management, all of the components of productivity come into play.

Announcements of productivity goals or productivity increases most likely are interpreted by your employees as a demand to work harder. This is simply not true. When introducing sidewalk snow management into your company's service offerings, you want the crews to work effectively and be accountable for the results of their efforts.

This is true not only from the standpoint of the contractor, but also from the viewpoint of the customer.

Crew Management Strategies

People are, by nature, territorial. We tend to accept responsibility and be more accountable when our territory is defined. Often it is difficult for the snow management laborer to accurately perceive the scope of "his or her territory," because in snow and ice management the areas are so large and diverse. It is also difficult for management to track performance.

Sidewalk snow work is a labor-intensive activity and most payroll dollars

Snow and ice on a sidewalk create an unsafe environment.

To be profitable, sidewalk snow removal requires attention to productivity levels.

are allocated to production staff as a percentage of the whole labor dollar outlay. The labor intensity of sidewalk snow management has improved and the percentages may change in the future, but people are the heart of the business.

Even with the introduction of new equipment that makes sidewalk snow management more efficient and affordable, labor is still a large part of the mix. Such advances in mechanized sidewalk clearing are best suited for large sites with extensive sidewalks. Relatively short walks will likely be the domain of the human laborer.

With the availability of reliable labor always being an up-and-down situation, it is likely the industry will continue to experience periodic labor shortages. Snow-and-ice management companies need to evaluate the importance of the production worker. It can be difficult to accept the reality that the crew member in snow management operations – often the lowest paid and generally an on-call, part-time employee – is the key element of a sidewalk snow management operation.

A production unit for sidewalk snow management consists of a crew that requires labor, equipment, material and transportation. The labor for one production unit usually consists of one crew leader or working foreman and one or more crew members.

The Crew Leader/Foreman. The crew leader has emerged as a specialist with expanded responsibilities for managing the snow and ice management on any specific site. The difficulty and expense associated with communication systems and direct supervision of mobile service crews, coupled with the need to have an experienced employee on the property at all times, has reshaped the value and job description of the traditional crew leader.

Companies that recognize the value of the crew leader's expanded role will streamline their organizational chart by eliminating middle managers and production supervisors. They will redistribute these responsibilities and upgrade the role of the crew leader.

The Production Worker. Most workers are attracted to the snow and ice management industry because they like to work outdoors, or they are doing it as part of a year-round activity that includes landscape or property management. Normally, they do not object to physically demanding work.

However, companies must provide employees with the proper tools to do their jobs since snow-related tasks are more physically demanding than landscape-maintenance, pavement-maintenance or equipment-operator work. Winter-season work may require a pay differential.

What specific traits should companies look for when hiring a sidewalk snow management crew member? Look for a person who needs to be active at all times and who appears to be bursting with energy. The crew member usually has little regard for detail and is difficult to train in a classroom setting. They learn by doing and are good candidates for on-the-job training.

In reality, most employees on sidewalk crews leave our industry because they do not see themselves managing a production operation. This is commonplace in other service industries, lawn care, landscape installation and pest control, as well as snow.

The Production Team. A sidewalk crew, large or small, should be staffed and organized so that the entire operation is a support system. Sidewalk work should be viewed as a production-oriented task and should be structured so management does not interfere with production. Crew workers perform best when they are managed as team members rather than laborers. They need specific goals set for each snow event.

The company standard – either your company's or the customer's – for performance must be demonstrated by the crew leader. The procedures that the standards are based on must be taught

while a snow event is taking place, thus putting more emphasis on the role of the crew leader and production workers.

Unfortunately, snow events in some areas of the country are few and far between, thus making the retention and teaching of proper-production principles difficult. Most people want more than a paycheck for a day's work, and part of a company's responsibility is to help build self-esteem in each member of the team.

Small Crew vs. Large Crew

The most efficient crew size for performing sidewalk work has been discussed, argued and subjected to trial-and-error testing. Since sidewalk snow management emerged as a viable part of the industry, the issue has become more important.

By adding mobile crews, companies discovered the importance of correct crew sizing. In a competitive labor environment, the need for higher productivity and increased quality suggests a new look at the sizing sidewalk crews.

Small-Crew Theory. Let's look back at our experience with working with small crews. At some point in our careers, most of us have worked as a one-person crew. Remember how much you could accomplish in one long day? Remember the first really good helper, the one who read your mind and did what you wanted them to do? You increased your production when you added the helper, but you did not double it.

Sidewalk snow management and snowplowing, for that matter, is a series of one-person tasks. Unlike construction crews, sidewalk crews do not handle heavy or awkward materials or equipment requiring more than one person to improve efficiency. This lack of synergistic benefit on a per-task basis encourages us to think of our crews as combinations of one-person crews.

Loading heavy sheets of plywood is a good example. One person can load 30 sheets per-hour by himself, but a crew of two can load 75 sheets in that same time. The difference is called synergy, which means the whole is greater than the sum of its parts. By working together, the plywood loading crew actually increased the output per person from 30 sheets per-hour to 37.5 sheets per-hour. What would the effect of adding a third person to the plywood loading crew? It would be detrimental to the overall production synergy.

In sidewalk snow management work, the opportunities to perform activities that have a positive synergistic effect do not exist. In fact, the opposite is true. When companies increase crew size, they lose efficiency.

Consider this example: A one-man crew goes to clear a sidewalk and it takes him four man-hours to complete the task. Send two people with the same equipment, traveling the same distance, to clear the same area, and the task is completed in 2.4 hours (or, 4.8 man-hours). The two-person crew did it faster – 2.4 hours vs. 4 hours for the one-person crew – but more time was spent in expended man-hours for the two person crew. Thus, it was less efficient in terms of man-hours (labor). That inefficiency translates into fewer profits due to higher labor costs.

Large-Crew Theory. In many cases, clearing a sidewalk in a little more than two hours, rather than half a day, is a worthwhile trade-off for the inefficiency, especially if there is a heavy-snow event taking place. The important issue is to recognize that the more people you send to do the job, the faster it is completed, but your efficiency is reduced. Your cost is proportionate to man-hours spent, not elapsed crew time.

One drawback of small two- or three-person crews on large properties is that they cannot complete the work fast enough. They spend too much time on-site or do not get the job done in a timely fashion. One answer to that problem is increasing the size of the crew. You should be able to send

TOP SNOWSTORMS IN UNITED STATES' HISTORY

#5 SUPER STORM OF 1993 – THE STORM OF THE CENTURY

March 12-13, 1993

This extremely intense and massive storm tracked from the western Gulf of Mexico to the Florida Panhandle and up the eastern seaboard to Massachusetts. Tremendous snow amounts covered a huge area, with many locations breaking snowfall records. Amounts ranged from one foot in southern Alabama to over 40 inches in the state of New York. In the mountains along the Tennessee/North Carolina border a whopping 60 inches fell. Winds of 70 mph were common across this large area with drifts to 20 feet high. Nearly 300 deaths were blamed on this storm.

Source: National Weather Service (www.crh.noaa.gov)

Small and large crews both have their advantages and disadvantages.

as many as six workers to one property, "knock it out," and then move on to the next site.

Large crews are fun to work with since they appeal to the social side of our nature, making it easy to build enthusiasm. Large crews make the members feel safe, secure and that there are enough of "us" to get the job done.

Crew leaders like large-sized crews because absenteeism does not cripple the production effort. Leaders, especially non-producing supervisors, like a lot of people to look after – it makes them feel needed. Crew members like large crews because they provide a team atmosphere and they have more freedom to do the things they enjoy as long as they remain productive. Property owners and managers like big crews because, in their mind, the more people working at their property the better. In fact, customers will often demand contractors "get more people" on the job and "get it done."

Crews working a specific route are often sized to service the largest prop-

erty and crews seem to grow almost by themselves. Supervisors and production managers have been known to want to add one member as insurance against anything going wrong. However, this is a sign of mismanagement, not efficiency. Everyone likes large crews except the person responsible for profit.

Large-Crew Myths. There are several myths involved with using large crews to clear sidewalks. The first is that large crews increase man-hour efficiency. Another popular myth is that large crews ensure quality work. This was born in the belief that it takes more time to do quality work and non-quality work is faster and saves time. Neither one of these is true. Quality is the result of a process that includes trained employees operating the correct equipment according to set procedures. In large crews where accountability is minimal, quality is often sacrificed.

Owners and managers like the feeling of having large crews attack a job, but this comes at a price. The first so-

139

A well-trained crew leader and a well-trained two- or three-person crew usually delivers the best results.

lution when you get behind is to add people. In fact, desperate property managers may attempt to dictate specific crew sizes and threaten to withhold payment if these demands are not met. In most cases this knock-it-out approach is an attempt to correct performance problems and force the contractor back on schedule. In this situation, don't increase the crew size.

The best strategy is to bring in a separate crew and divide the property into appropriate zones. Once back on schedule, the customer will become accustomed to, and accept, fewer people on the site during a snow and ice event.

The myth that one strong crew leader can supervise five people as easily as two is false. Some leaders will try to keep a specific crew together because they claim they are easier to supervise, but in reality the "herd mentality" further reduces productivity. A large-crew supervisor must make a choice to reduce their own productivity to keep five men up to speed, or allow their productivity to drop to maintain individual productivity. Most large-crew supervisors do a

little bit of both and lose both productivity and quality in the process.

The combination that seems to work best is a full-time crew leader with one or two well-trained crew members who require less direct supervision. Divide large crews into smaller two- and three-man crews and teach them to function as separate units.

For example, when a large property requires more man hours than a three-man crew can generate, divide the property into two zones and send two crews to do the work. It will be less expensive for the customer in the long run and you will look much better at budget-review time.

Each two- or three-person crew should have production and quality goals for each snow event. Even though they may be in competition on the same property, they should be evaluated on that particular snow event's performance.

Sidewalk Snow Removal Equipment

Putting shovels into the hands of crew

members on sidewalk duty is a workers compensation claim waiting to happen. Hand pushers are much more efficient and safer for the worker to use. One cannot actually lift up snow, but can quickly push the snow off to the side.

There are new versions of the stick plow on the market that contractors can consider purchasing. The Dakota Snow Blade looks as if it won't work well, however experienced snow contractors swear by these units since they are highly efficient in snowfalls of less than 4 inches.

The use of four-wheel ATVs for sidewalk work has become a popular option for contractors. These units are generally affordable, but they are also inherently unsafe and susceptible to employee misuse. The biggest drawback comes if a plow-equipped ATV encounters a sidewalk imperfection. This can cause the operator to be ejected off the front end of the unit, leading to serious injury and even death.

It's also important to note that at this book's publishing (June 2011), there are no spreaders built for efficient use on four-wheel vehicles.

Mechanized equipment has advanced considerably over the past few years. It used to be the only really productive mechanized units were essentially designed for municipal and airport snow management. However, these units were expensive ($150,000+) and usually out of the realm of possibility for commercial snow contractors. These units move a lot of snow quickly.

There are 4WD articulating units available for less than $35,000, with attachments, that perform very well for snow contractors on condominium, HOA and retail sites. These units are designed with safety in mind, feature heated cabs and four-point safety harnesses for operators. These units have deicing spreaders designed specifically for use by contractors doing sidewalk work.

Sidewalk snow removal is a labor-intensive job for contractors.

141

Chapter Fourteen
WORKING WITH DEICING PRODUCTS

Chapter Highlights

- Deicing Materials
- Application Strategies
- Pricing Deicing Services
- Deicing Sidewalks
- Alternative Deicing Materials

Customers will often expect more of you than simply moving snow from their parking lots, sidewalks and other walkways. Most customers, especially commercial and industrial clients, expect that you will also keep their facilities free of ice. Liability from ice-related falls or accidents pose significant risks for commercial clients, so they often require you to keep their location free of both snow and ice.

This is a good thing for your company. Deicing is one of the most profitable services your company can offer, and it is relatively easy to integrate into your plowing operations. Ice-control margins peaked in the late 1990s at 75%, but with the economic downturn and downward-pricing pressures, margins at the time of this book's printing (June 2011) have fallen to around 60% to 65%. This is still, however, a significant profit margin.

The use of deicing products within the framework of the snowplowing industry is a huge topic and could warrant an entire book in and of itself. What we don't know about controlling ice buildup on pavement dramatically dwarfs what we do know. The study of ice-control practices goes on daily by various state departments of transportation (DOT) that operate in cold-weather climates. The dollars allocated for such controls runs into the billions each season.

With the public's expectation to be able to travel paved surfaces safely during the winter and with the ultimate goal being bare pavement, government agencies are learning what is needed to fulfill these expectations.

Private snow management contractors have also been driven to become more knowledgeable about ice-control practices. As contractors become more well versed in the proper way to manage their ice-control business, they have come to realize just throwing down more salt is not profitable, not environmentally friendly and is simply a bad business strategy.

Deicing Materials

For decades, rock salt was the deicer of choice. The supply was plentiful, the product was effective at moderately cold temperatures and it was a very inexpensive product to purchase. The harsh winter of 1992 led to salt shortages in the Northeastern portion of the United States as demand far exceeded supply. Pricing structures rose in response to the demand and municipalities and states confiscated available salt supplies, which further aggravated the situation.

Alternative deicing products were developed in the aftermath of this salt shortage and considerable experimentation took place with these alternative products. Scientific methods were introduced into research studies and the push was on, led by state DOTs, to develop ways to safeguard the public even in the face of a shortage of available product.

Private and public companies poured millions into development of a number of alternative deicing agents, as well. Chlorides and chloride mixtures were formulated and tested, and one company went so far as to test – with some degree of success – the use of distilled corn solids to treat slippery pavement conditions.

The latter part of the 1990s gave North America relatively mild winters and salt supplies were, for the most part, adequate. However, as contractors became more sophisticated in servicing customers, the demand for alternative products increased.

The winter of 2000-2001 saw a radical shift in the weather patterns over North America that brought back the old-fashioned winters we had become accustomed to. The following season contractors, municipalities, states and other public entities substantially increased rock salt requests, creating an artificial salt shortage. This drove up pricing again for winters thereafter.

Given that winters, generally speaking, have stayed active with heavy snowfalls continuing to occur, combined with the renewed environmental interest in how we control slippery conditions, the threat of not being able to properly service public and private customers caused renewed interest in alternative methods of controlling ice buildup on paved surfaces.

The occasional shortage of salt, combined with the desire of DOTs and snow contractors to find faster and more efficient ice-melting materials, has forced the creation of several relatively new technologies in deicing. These include development and use of calcium and magnesium chloride, calcium magnesium acetate and a whole host of environmentally sensitive products.

Manufacturers have also developed liquid chemicals that can be applied

---

TOP SNOWSTORMS IN UNITED STATES' HISTORY

#6 BLIZZARD OF 2005

Jan. 20-24, 2005

What started out as a clipper system diving into the northern Plains quickly turned into an unusual type of blizzard as it hit the Lower Great Lakes. Record snowfall blanketed much of the Upper Midwest, northern Ohio Valley, and most of New England. Snowfall totals ranged from 5 to 13 inches across the Midwest and 8 to 37 inches across southern New England. Areas around Boston reported snowfall rates of 3 to 5 inches per hour for a period of time. One nearby city recorded 7 inches of snow in 75 minutes. Boston officially recorded 22.5 inches, which contributed largely to breaking an all-time record for the amount of monthly snowfall (43.3 inches) for any month.

Source: National Weather Service (www.crh.noaa.gov)

Removing ice from parking lots, sidewalks and other traffic areas can be a very profitable service.

before a storm to prevent ice from forming on pavement. This is called anti-icing.

Techniques and products have also been developed to treat bulk rock salt with a liquid substance that lowers the temperatures at which salt is effective at fighting ice, while at the same time reducing salt's corrosive nature. These non-traditional materials and techniques are gaining popularity and are worthy of further investigation.

Additives developed and sprayed onto rock salt were also developed, with varying degrees of success. One company developed an outstanding marketing strategy to promote their product. As these products came on-line, costs varied from $8 per ton to $16 per ton for the additive alone. When rock salt pricing was between $30 and $50 per ton, the percentage increase for using these products – in conjunction with standard sodium chloride – was fairly substantial.

When the cost to the contractor increases upwards of 25%, it greatly affects profit. However, as the cost of rock salt escalated, using the additive

products became more affordable – at least from a percentage-of-cost standpoint. As a result, a few of these products flourished in the marketplace and while they do work, to what extent is still under debate.

The most common deicing technique among snow contractors is still the use of rock salt applied through pickup truck mounted spreaders or through V-box spreaders mounted on dump trucks. As a result, most of the discussion in this chapter will be dedicated to traditional applications of deicing with rock salt.

Why Does Salt Melt? Rock salt, in and of itself, will not melt anything. This is why we can store rock salt in quantity. Rock salt does something when provided with a catalyst – moisture. In many cases, this moisture arrives in the form of snow or ice. Once moisture is introduced into the equation, salt goes to solution, meaning it dissolves into a liquid brine (23%, by weight, salt).

The brine spreads under the ice material and breaks the bond between the ice buildup and the pavement surface. This is how ice is separated from a road

Truck-mounted spreaders are commonly used by contractors to apply deicing materials to a variety of accounts.

or parking lot surface. The ice can then easily be removed from the pavement with plowing equipment.

Because of brine's chemical properties, it also lowers the temperature at which the liquid will freeze by lowering the vapor pressure of the solution, thus reducing the surrounding ice to water. With a lower vapor pressure in the solution, the solid ice wants to go into solution and become a liquid. As a result, the ice melts.

Heat is also required to allow salt to go into solution. Salt is effective at drawing heat from the air and pavement.

Application Strategies

Ice control on parking lots requires some capital investment. While a truck is needed, it does not necessarily need to be a truck outfitted with a plow. In fact, there are some compelling arguments that having a truck plow *and* salt is not an effective use of resources.

The initial venture into deicing services begins with the acquisition of a tailgate-mounted, electrically driven spreader holding 700 to 900 lbs. of deicing material. These units are not terribly expensive and allow the small operator to enter the deicing business with a minimal capital investment – usually less than $2,000.

The downside to this method of deicing parking lots and driveways is the deicing material must be manually loaded into the hopper, thus requiring someone to work outside in the elements several times during a typical evening plowing.

Generally, the material is loaded onto the truck in 50- or 80-lb. bags, manually opened and then dumped into the hopper. While inefficient from a labor standpoint, margins are still much higher than plowing.

The next step in building your deicing service is to purchase and mount a slide-in, V-box spreader. These spreaders are hydraulically or electrically driven units that hold from 1.5 to 7 tons of material. Some manufacturers make units that fit nicely on a three-quarter ton pickup, but care must be taken not to overload the hopper. The additional weight can cause problems with the truck's suspension, causing damage to the undercarriage and axles of the vehicle.

Once the spreader is mounted on the truck, whether on the tailgate or as a V-box, it is filled with salt and the mechanism turned on, spreading the salt while the truck drives over the pavement.

Most spreaders are equipped with mechanisms that allow the operator to control the amount of salt being applied. Once the operator determines how wide the salt path is, he or she can determine how many passes are required to coat the entire lot with salt or deicing product.

As your company continues to grow, using spreaders mounted on the rear boom of a backhoe can also reduce labor costs since a single operator with a pusher on the front of a backhoe can clear and deice the lot simultaneously.

Anti-Icing. Deicing is the most common means of removing ice from a customer's facility or residence. Deicing is the application of an ice-melting material *after* ice has formed on the pavement, sidewalk or other traffic area.

However, you can apply salt or an alternative deicing product prior to the snow or ice event. This technique is called anti-icing and it has grown in popularity, especially among contractors using liquid deicing materials.

Salt, sitting on the paved surface, is inert unless moisture is introduced and comes in contact with the granular rock salt. Once it starts to snow, the moisture causes the salt to dissolve into solution. The resulting salt brine prevents ice and snow from bonding with the pavement surface. Since no bonding takes place, once plowing operations commence the snow or slush is easily removed. This leaves a cleaner surface than if you

plow the site after the snow and ice has bonded to the pavement.

The nice thing is you can achieve this result by using only one-third the amount of product required for traditional deicing. If it does not continue to snow after completion of plowing operations, there is often no need to reapply salt to the cleared surface. If an additional application of salt is required, desired results can be achieved with considerably less material than you would have needed had you not been proactive.

All things considered, the astute contractor can actually use half the normal amount of salt by having a pre-salting program in place. Most contractors who pre-salt also make a very light application of salt after the plowing has been completed.

How Much Is Enough? When discussion occurs about the use of rock salt (sodium chloride) and its distribution to the pavement surface, there are always arguments as to how much product needs to be applied to the pavement surface to achieve the desired results.

Don Walker from the University of Wisconsin, one of the leading authorities on deicing in the country, stated that 200 lbs. of rock salt applied evenly on 1 acre of surface is adequate to reduce a light to medium buildup of ice to a liquid form.

Various DOT studies indicate that in a light-icing situation, 200 to 250 lbs. of rock salt per acre is all that is required to reduce a light accumulation of ice to water at approximately 28 degrees F. Under these conditions, the melting process will take 45 to 60 minutes to complete. A heavy accumulation of ice may require as much as 350 lbs. of rock salt per acre. This may seem absurdly low, but these low application levels are attainable.

In fact, recent studies indicate as little as 75 lbs. of rock salt will address a light icing on 1 acre of pavement. Unfortunately for contractors, the V-box, slide-in spreaders can only be calibrated down to about 300 lbs. per-acre distribution. There are spreaders on the market that can go as low as 75 lbs. per acre, but the cost of these units is well over $50,000. Normally, this is out of a commercial plowing contractor's price range.

Contractors often charge for half-ton and full-ton applications of rock salt for an acre of parking surface. More often than not, contractors apply more salt than is necessary. This has some negative consequences, including reduction of profits, the potential to overcharge customers and doing more harm to the environment than is necessary. The next time you see a white parking lot the day after a snow storm, it is likely the contractor over applied salt to the lot.

Pricing Deicing Services

Similar to snowplowing, deicing services can be priced many different ways. Per-pound and per-ton methods are the most common. While easy to administer and track for billing purposes, such pricing lends itself to the overuse of salt and other deicing materials.

The contractor often fails to realize this is happening, and he has little incentive to learn the correct application methods. While this statement may offend some veteran contractors, the facts show it is true. Unfortunately, some of this is customer-driven abuse and not necessarily the contractor's fault.

Customers often request by-the-ton or by-the-pound pricing since it is easy for them to determine what, in their estimation, the best price is for the job

at hand. What customers don't realize is that accurately tracking how much product is actually applied is not easily done, unless the contractor has thousands of dollars to spend on expensive monitoring equipment. Most contractors make an educated guess at how much product has been applied.

Unfortunately, some contractors use this fact to their advantage by artificially inflating the quantity of applied deicing product to ensure all the product is accounted for in the billing cycle. There is no incentive for contractors to establish good deicing practices based on the available facts. The customer is paying by the ton, the thinking goes, so use as much as possible to achieve the desired results.

Another way the customer encourages this abuse is their perception of what takes place during the deicing process. Society is impatient – they want immediate results. Add to this a customer's lack of understanding as to how the deicing process takes place, and it is easy to see why there is a likelihood of deicing material abuse and overuse.

While over application marginally speeds up the melting process, it is a huge waste of product. Salt will only stay in solution up to 23% (by weight) of salt in liquid. More than that and the salt will solidify and settle to the bottom of the puddle. This is why parking lots get a white tinge to them the day after the storm has passed. That white tinge is actually rock salt that has solidified and settled to the surface of the pavement area.

Two factors also lead to overuse of salt product. The first is customer-driven as he or she wants the ice gone immediately. If they step onto the property and see that the ice is not gone, they want to see material on the ice to prove you have been there. Feeling crunchy salt under their feet makes them feel comfortable, even if most of it is wasted product.

Keep in mind, rock salt will work at temperatures slightly below 0 degrees F. Unfortunately, the process slows to a point where you may have to wait all day for the melting process to be completed. In most cases, the customer finds this unacceptable. Applying additional product speeds up the process and gives the customer a feeling of security, but in reality, it just wastes material.

Remember, perception is reality to most customers. As such, contractors tend to overreact by applying more material. And, if the customer pays by the ton, then the customer is paying for the privilege of feeling good.

The second factor that leads to over-application of rock salt is employee driven. Most employees want to do a good job. They don't get up in the morning with the specific intent of doing a poor job. While it might appear this is the case, human behavioral studies indicate most people want to do well in their jobs.

When the salt-truck operator leaves a site, he wants to know he has done well. Knowing that the melting process can take as long as an hour after application, it is very difficult for the operator to achieve a level of satisfaction by leaving the site knowing ice or snow is still present. Thus, the employee needs reinforcement that the process is happening and that they have done a good job. Like the customer, they also want to see results.

The desire to see results might tempt the employee to apply more material in an effort to speed up the ice-melting process. Here again, education and knowledge are the key factors to a successful ice-control program. Even if you properly trained the operator on how much material is required to achieve acceptable results, it will go against the grain of human nature.

Giving the operator the visual confirmation of a job well done can be accomplished without excessive product application. One suggestion is to route the operator in a star or crisscross pattern forcing them to drive by sites

Deicing materials vary widely from rock salt to liquid deicers. This salt pile is being treated with a liquid additive.

they've already treated. Inevitably, they will visually inspect the treated sites and this will provide the positive reinforcement that the job was done properly. Granted, this may not be possible in all instances, but even one positive reinforcement is better than none.

Interestingly enough, this same premise works well with sidewalk crews required to apply melt ice after clearing snow. The minor inefficiencies created by not designing routes in a straight line will be offset by the dramatic reduction in material needed and labor spent at the account.

"Ah-ha," says the profit-minded contractor. "I'm charging per-pound for the material and per-hour for my labor and you have just reduced the amount of labor required, as well as the amount of material I can charge my customer for." Your ah-ha moment is true.

However, it should become more apparent to the profit-minded contractor that margins can be increased dramatically if the customer was charged per occurrence or per application.

For argument's sake, let's assume you are charging per-pound for material and per-hour for labor. It should be easy to compute what the customer has paid per visit to perform deicing services. Knowing full well you are over applying material, and that you probably can reduce the material cost by a significant amount through education and proper training, why not figure out what the customer has been paying on average per visit?

Then go to the customer and show them what they are paying you. During the meeting, offer a per-occurrence or per-application pricing strategy that lowers the customer's total per-visit cost by as much as 20%. Everybody wins in this arrangement.

The customer pays less and respects you for reducing their cost and you increase your margins by reducing the amount of material required and the labor needed.

The key to all of this is education and

Contractors should experiment with different deicing strategies and chemicals to determine the best product for each situation.

good management practices. Operators need to be properly routed and trained as to how much product is necessary to achieve the desired results. As a manager, you have to monitor what they are doing to be certain everyone is performing as they should and applying the appropriate amount of material.

Let's take this thinking a step further. If you are charging a customer a flat, seasonal fee that includes salt applications, there are other things you can do to increase your margins and deliver superior customer service.

Under the seasonal-price strategy, contractors can increase margins by pre-salting parking lots. Often referred to as anti-icing, this is one way to dramatically reduce material costs normally associated with ice control. It also gives the customer a better end-result.

Again, we can hear the traditional contractor saying this is a hard sell to customers. And, for those who charge per occurrence or per pound, this is true. It is very difficult in the private sector to convince property managers and owners this is a wise use of their money. Getting a property owner to accept that you are going to salt his or her lot before the storm – and possibly when the sun is shining – can be difficult.

However, on those accounts where customers pay one fee for the entire season, it is in the contractor's best interest to utilize the most efficient and cost-effective methods available. When that happens, customer service levels improve while the contractor's margins increase.

Deicing Sidewalks

There are numerous products on the market geared towards sidewalk deicing and almost all are blends of various products. Normally, rock salt is not the deicer of choice for sidewalks and there are valid reasons for this. Some contractors are of the opinion that salt causes "spalling" on sidewalks. This is not true. When using rock salt on sidewalks, you are lowering water's freezing point. Essentially, the snow and ice on the sidewalks is reduced to water.

Sidewalk concrete is not as strong as road-surface concrete because the weight placed on walkways is considerably less than on roads and bridges. . Concrete is porous. And with the concrete mixes used on sidewalks, the top layer is pitted with voids. You may not see them with the naked eye, but they are there.

Water, which results following the application of rock salt, permeates the concrete and fills those voids. When the temperatures drop at night the water refreezes, expands and "pops" the concrete up. This is known as spalling. Rock salt does not causes the spalling. Rather, the water that remains on the concrete surface is the culprit.

Some customers specify the use of calcium chloride on sidewalks, believing it does not cause spalling. This, too, is not completely true. Calcium chloride reacts violently with water and generates a lot of heat. In fact, if you place your hand in a bucket of water, and then thrust it into a bag of calcium chloride, you can severely burn your hand from the resulting chemical reaction.

This reaction generates so much heat that evaporation results. The air in winter is normally very dry. When heat is added to the equation, the water evaporates quickly. As a result, when people see dry pavement after an application of calcium chloride they believe the calcium absorbed the water. This is not the case. Since the water does not sit on the concrete surface, it does not have the

SNOW FACT:

New York State is home to some of the snowiest cities in the country with Syracuse, 115 inches of snow per year, and Rochester, 93 inches per year.

Source: National Climatic Data Center

To avoid burns, contractors should use gloves when handling calcium chloride-based products.

opportunity to fill the voids in the concrete's top layer and does not refreeze at night. Thus, spalling is prohibited.

Most bagged sidewalk deicing products sold today are "blends." These blends are made up of sodium chloride, calcium chloride, magnesium chloride, potassium chloride and calcium magnesium acetate. The percentages change depending upon who is blending the product for packaging. The main reason for blending the products is to control cost. Rock salt is cheap even with the escalation of pricing over the past decade. Other products are more expensive and blending keeps costs down and makes the products more affordable.

I'm often asked, "What is the best product (or blend) to use on sidewalks?" Unfortunately, the most honest the answer is, "The best product is the one your supplier has on the shelf the day you walk in." I cannot imagine a retailer telling a potential customer, "Don't buy this product. Go down the street to XYZ Store and get theirs, it's better." Contractors need to be educated about the various products available for use as deicing or anti-icing applications.

Applying Deicing Products. There are various methods for applying product to sidewalks. Some contractors use a 5-gallon pail, partially fill it with material and then walk the sidewalk with the pail under their arm using the "chicken-scratch" method of distributing the product by hand. When using a calcium chloride-based product, doing this without gloves is quite dangerous as the result is painful burns to the hand.

Other contractors use rotary, fertilizer-style spreaders. This often results in application of deicing products to shrub beds, car bumpers and onto areas where deicing is unnecessary. Rotary spreaders come in plastic and stainless steel. When asked which is better, I usually responds that the $40 plastic spreaders fit into the dumpster at the end of the season just as well as the $350 stainless-steel spreader. All of the deicing products on the market can be detrimental to the workings of these type spreaders.

Drop spreaders work very well for applying deicing product since the product is distributed exactly where you want. However, this type of spreader has the same issues as the rotary spreader in terms of the useful life of the unit. This author has found that the only really good drop spreader on the market is the Epoke Epomini 5. This unit is coated with epoxy that reduces the corrosion factor brought on by the chlorides present in the various deicing products. Of course, with such durability comes increased cost. This unit costs in excess of $500 at the time this book was published (June 2011).

Using Sand And Salt. Let's take a look at the use, or rather misuse, of sand as a traction agent. Sand as a traction material has a very short life span of effectiveness. This product is erroneously used in markets under the guise of being an adequate way to prevent cars from sliding on ice. This is not true. The effectiveness of sand on roadway surfaces as a traction control product is poor at best.

Iowa and Idaho DOT studies have shown that sand provides effective traction for no more than 10 vehicles before losing all traction ability. However, in certain markets, contractors and customers insist on using predominantly sand mixtures for deicing and traction control on parking-lot surfaces. The progressive contractor who knows the facts will shy away from using sand for deicing in lieu of materials that are better suited for the job.

The inevitable cleanup is another reason sand is not a good material to use as a deicing and/or traction agent. Sand clogs storm sewers and collects along curb. Picking up sand material in spring can be costly to the customer. While a source of revenue for some contractors, it is not in the customer's best interest.

Alternative Deicing Materials

As noted earlier in this chapter, salt is still the most common deicing material used in the market. However, alternative deicing materials have gained credibility among contractors in recent years. Liquid products, though more expensive than salt, both in material cost and the cost to apply them, can be highly effective if used properly.

Since they melt ice faster and have a longer residual life than salt, these new deicing chemicals can boost the efficiency of a contractor's ice-melting operations. There are also several organic-based deicers available that can be used in various environmentally sensitive situations or when the customer requests such a product.

Unfortunately, lack of knowledge impedes the use of alternatives to salt more than a lack of the products' performance. All of the liquid materials have very specific uses, temperature ranges and usage requirements. If not used as directed, performance levels fall and customer satisfaction drops.

NOTES:

Chapter Fifteen
MANAGING A SNOW EVENT

Chapter Highlights

- Preparing For The Storm
- During The Storm
- Dealing With An Extreme Snow Event

The most important part of any snow- or ice-management plan is to be ready, willing and able to go to work when it snows. Implementation of the overall plan is crucial to success. What good is it to have all of the business aspects covered, all of the equipment ready and a plan in place only to fail when the snow comes? Let's take a step-by-step approach to how contractors should approach an actual event.

Preparing For The Storm

There are several key steps a snow contractor will want to review before each and every snow or ice storm hits his service area. Reviewing these steps before each storm helps you to better organize your efforts and manage your crews' efforts.

Weather Monitoring. The snow-and-ice management process starts with the advanced prediction of a snow or ice event. Local forecasters usually give an advanced warning at least three days prior to the event. Keep in mind, though, that television forecasters are predicting for the masses. The astute contractor will not use this general forecast as gospel.

However, it is an advanced warning of what might be ahead. Many contractors religiously monitor The

Weather Channel during the winter season and some even subscribe to satellite weather services, that provide on-call or Internet-based, 24-hour weather prognostication and surveillance. These alerts can even be sent to cell phones or other mobile devices. Smart phones allow you to view the local radar at a moment's notice.

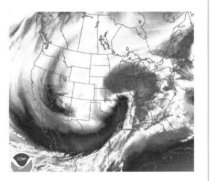

Being proactive when preparing for storms will help your company and employees be more productive.

Review Equipment. Three days prior to the event, a review of the snow-management equipment available to you is advisable. The equipment should already have been tested and "winterized." However, one last look will assure that no surprises are in store when the snow falls. If you have gone several weeks since the last storm, making sure the equipment is in working order can save considerable headaches when the snow is flying.

After checking the equipment, it is

also advisable to contact your subcontractors or those employees who will be out in the elements fighting the storm. Sometimes they are not paying as much attention to the weather as you are, and a short, heads-up phone call saves considerable confusion when you call them out at midnight a few days down the road.

Working The Phones. As the actual time for the snow event gets closer, an additional phone call, text or email to subcontractors and employees before the predicted event will tell you if you have any surprises coming. A sick employee can be replaced with a back-up person if you are aware of the illness ahead of time. The same goes for the subcontractor who might be on vacation, out of town or just plain not ready to go.

An additional phone call, text or email during the daytime hours – assuming a nighttime storm – will alert employees you will be calling. The, "I didn't get the message," excuse can be eliminated if they know in advance the call is coming. After the contact is made, then supply the dispatcher with

the list of who is not available. This can save considerable time tracking down available personnel at the beginning of the plowing event.

Usually, the person who makes the decision to "go" will not get much sleep the night prior to the storm's arrival. Constant attention to the local forecasts, a satellite feed of current radar and what's actually happening outside takes up most of this person's time.

During The Storm

Once a snow or ice storm arrives, your life can be a little hectic. The fast-paced nature of this industry requires you remain focused on your objectives – delivering superior service to your customers. Assuming you have done the proper pre-storm preparation, here are some important tasks you will want to keep on top of during a storm.

Monitor Your Customers. When it starts snowing, the decision maker will often drive around to see the conditions in the market area. They may gain input from others who are on alert to watch the conditions. In areas where weather can be different within the same geographic service area, input is usually necessary to ascertain how many crews to call out to work.

When the trigger depth is reached in a specific service area, the decision to plow is made. Phone calls to the dispatcher and the area supervisors should be made at this time, usually from the decision-maker's vehicle via cell phone. The dispatch team arrives at the office, area supervisors call out their crews and crew leaders call out their operators. Ideally, the entire call-out process is completed within an hour of the decision to "go."

Effective communication with employees and customers during a snow event is good practice.

156

Pre-salting is the application of deicing material before a storm begins.

Dispatching. Crews should have preordained routes to cover and these routes are usually geographically clustered to eliminate undo travel between sites. Once the call is made, crews proceed to those route areas and begin service. Remember, proper and efficient routing is essential to the successful completion of the plowing operation.

The dispatcher begins communicating with area supervisors to determine where the company is short on drivers, equipment operators or subcontractors. The dispatcher communicates with customers as needed. Continual communication between the dispatcher and area supervisors keeps everyone informed of the conditions in the field, the progress of plowing operations and any unusual events, including accidents, breakdowns and changes in storm intensity.

The dispatcher records all of these benchmarks in chronological order either via digital recorders, noting everything on paper or on a computer spreadsheet. Often, the area supervisors carry digital recorders to record their own observations. Some cell phones also have voice-recording technology built in. The notes are then transcribed after the event and placed in a file for future reference, just in case. Remember, proper record keeping is essential to a snowplow operation since it allows for proper billing, customer communication and liability protection in the event of an accident.

Assuming that plowing operations started about midnight, all area supervisors report to dispatch by 5 a.m. to update the office as to what is happening. Hopefully, all is well and the crews are proceeding in an orderly fashion to clear snow at all assigned locations. However, in the unlikely event that some things did not go as planned, this is a good time to alert everyone where problems exist and what is being done to expedite changes in schedules to accommodate the problem areas.

The dispatcher has an overall view of the progress and can assist in making decisions as to where to reroute a crew or a plow to shore up the area that may be behind schedule. The area supervisor is responsible for keeping track of what has been plowed and what has not yet been completed. The dispatch crew is responsible for knowing the status of the various areas as far as the number of completed accounts.

Today, technology has allowed many companies to eliminate the dispatcher position. Employees and subcontractors can now report progress through interactive voice recognition (IVR) software or even an iPhone app. All data input is done with a computer that talks with the operator and records what has been accomplished. Whoever is in charge can monitor, in real-time, the progress being made in the field through a computer that is often mounted in a truck. Real-time reports can tell if one crew is behind schedule and might require assistance to complete their assigned sites within the allotted time frame.

After the event, the same software can generate a subcontractor's invoice for services provided. This same software can also report to the accounting package, without human interaction, what has been completed so that customer invoices can be immediately produced and emailed. Profit or loss reports are available before "the boss" gets back to the office. Invoices can be immediately emailed to the accounts-payable department at customers' locations. If the event ends by 6 a.m. and removal work is completed by 8 a.m., it is *possible* to have all paperwork – inbound invoices from subs, outbound invoices to customers and P&L reports generated – completed within minutes of the field work concluding.

Salt Runs. If salt trucks are dedicated and not equipped with plows, then these units usually start after the plowing crews have begun their operations. It rarely does any good to salt the lot just prior to plowing operations. If a company does anti-icing – the application of deicing material before a storm hits – it is done hours before the start of the snow event. If the plow trucks carry deicing equipment such as tailgate spreaders or V-box spreaders, deicing

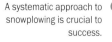

A systematic approach to snowplowing is crucial to success.

operations generally take place immediately after completion of the plowing operation. Proper scheduling of the deicing operation can greatly enhance the look of the finished product.

Quality Control. Area supervisors and crew leaders are responsible for completion of plowing operations. They are also responsible for the quality of work their crews are performing.

A high-quality job is one where the customer looks at the site after plowing operations are completed and is satisfied with what he or she sees. Corners are square and tight with no snow feathering out into the lot. The plow lines are straight and not wavy. Light poles don't have unsightly snow piles haphazardly left behind and curb lines are evident and piles are tight in specified snow-dump areas.

Fire hydrant locations can easily be identified and snow is not piled upon sidewalks, left in driveway aprons, in the street or drive lanes. A quality job is easily identified but so is a poorly done job. Some customers do not require "high-quality jobs" and a business decision has to be made whether to deal with this type of customer.

Paperwork. At the onset of your endeavor into the snow-and-ice-management business, it is unlikely you will be using automatic reporting, invoicing and accounting technology. As the storm winds down, area supervisors are checking sites for quality control and touching up lots to make sure all looks good to the customer when they arrive. The supervisors then turn in hours and completed plowing data to the company's dispatcher who checks this information against his own in-house information-gathering system to ensure accuracy. Ev-

In addition to making sure subcontractors are ready to go, the dispatcher should call out sidewalk crews when needed.

eryone knows the job is not done until the paperwork is completed.

Subcontractors will call in to check that their hours were properly turned in and to be certain that their records match those of the company. This task becomes unnecessary if computer and IVR technology are in place. In lieu of that, the benefit from the double-check system is that everyone is aware that the system ensures all parties will be paid the correct amount at the correct time.

In the case of ongoing snowstorms, these checks and balances are necessary because operators can often get confused if plowing every day for several days in a row.

Clean-Up Work. The next night, crews may be called out again to perform clean-up work. Packed snow gets mushy, requiring crews to clean off slush in parking lots to keep them safe. Snow piles may have to be relocated to accommodate customer needs and often a reapplication of deicing materials is required to melt off any remaining ice. Area supervisors can make these decisions in conjunction with the dispatch team.

Extreme snow events will fully test your abilities and company-wide systems.

A word of caution – many companies attempt to utilize speed crews for clean-up work. The speed crew is usually made up of experienced snowplowers who know the sites. It is thought that these plowers will get through the sites quickly if the snow accumulation is light or if clean-up work is all that is necessary. Seasoned snowplowing companies indicate that this may not be the most effective way to go. Those not on the "speed crews" are often offended that they were not utilized for the "easy work." Those on the speed crews are hard pressed to get the work done on time as undue pressure is placed on them to perform quickly. Quality is sacrificed in lieu of time constraints and the customer can become dissatisfied with the finished product.

Veteran contractors often dispatch the entire area crew to do the clean-up work. That way, everyone on the crew learns what is expected in the clean-up process and they get some of the easy work instead of feeling they were hired only to beat up their equipment during bad storms.

In the long run, these operators are grateful for the trust placed in them to do an efficient job during the clean-up operation and will work faster knowing they will get called out the next time there is an easy night coming.

Follow-Up. Placing follow-up calls is the last order of business when working a snow event. This is the most important part of the plowing operation because it is much easier to retain satisfied customers than to find new clients. And, if there is a problem, it is best to find out about it immediately rather than at contract-renewal time when little can be done to rectify the customer's concerns.

Dealing With An Extreme Snow Event

Every once in awhile, a snowfall occurs that is truly exceptional for the geographical market area. For some markets, this type of storm might occur only once a decade, while other regions may experience one extra-heavy storm per season. In some markets, the catastrophic event might be 6 inches of snow overnight and in other markets such an event might be defined as 3 feet of snow in a 24-hour period.

In any case, these extreme snow events must be dealt with differently. Back-up equipment is often necessary, and this equipment should be lined up prior to the season. Such avenues for back-up equipment might be as simple as having the home phone number for an equipment-rental outfit that can supply you with loaders in an emergency.

Keep a list of plow drivers who are willing to work for you if a bad storm arises. These might be people who have small routes of their own and would be willing to work for you to gain extra income. Some companies have names and numbers of plowers from as far as two-hours away, who can be recruited in the case of a monster snowfall.

Plans should be in place to relieve plowers and operators who are involved in an ongoing storm situation. While constant plowing during the storm is a good way to keep up with monster snowfalls, it must be kept in mind that a fatigued loader operator can do considerable damage to property if not given time to sleep. Often, plowers only need time to shower and eat to continue working over the long haul.

Damages occur in increasing frequency as plowers become fatigued by an ongoing storm. It does a company little good to have a subcontractor damage his own equipment and not be available to continue plowing. Thus, it is often much cheaper to let subcontractors and exhausted employees go home for a few hours, rather than to lose them for a few weeks.

Record keeping and customer communication become increasingly more important when a catastrophic storm occurs. Supervisors and dispatchers need to communicate often so that proper documentation takes place during the storm.

Customers will call stating, "You haven't been here yet," and you need to know whether that is accurate. It is difficult to explain to a customer when they receive the invoice that you were there two or three times providing services when they are certain you were not there. This often happens because the customer looks out the window and sees "white" but doesn't see the piles you made during a previous visit.

Sometimes a simple call, text or email to a customer can alleviate this misconception. If all incoming calls for additional service are routed to the dispatch center, and assuming the dispatcher has the information, then the customer can be told, "We were there at 4 a.m., and we can return to re-plow if you want." In this scenario, the customer is informed that service has been provided and there is no confusion when the invoice arrives.

NOTES:

Chapter Sixteen
CASE STUDIES

Chapter Highlights

- ´Twas The Ice Before Christmas
- One Slip and You Can Take The Fall
- A Weighty, Costly Issue

Snow and ice management provides contractors with unique challenges.

CASE STUDY 1:
'TWAS THE ICE BEFORE CHRISTMAS

BACKGROUND: An established snow operation – we'll call it ABC Snow Contracting Co. – handled several sites in a Midwestern city, including a homeowners' association (HOA). The contract with this specific HOA calls for ABC to plow and salt all the driveways within the association. The company is not responsible for the street plowing, unless specifically requested.

They've serviced the contract for a couple years, having been awarded the business based on their ability to perform, as well as on referrals from two other local HOA's who they've worked with. To this point, they have been doing a good job and received complimentary letters from various individual homeowners. ABC is charging per push for plowing and per application for deicing. They deice with rock salt (on the asphalt driveways) and a blended product on the walks.

The season starts out with a bang. By Christmas the area has experienced almost their entire seasonal average in one month. On Dec. 18 a phone conversation takes place between the property management firm overseeing the HOA's maintenance and a representative from ABC Co.

ABC stated they were verbally instructed to "cut back" on the amount of service because the HOA is overbudget for snow maintenance and is worried they won't have enough to cover the remainder of the season. During this telephone conversation, the service level is upped from zero tolerance to a 2-inch trigger; deicing is placed "on call" instead of being automatic.

No follow-up conversations took place and, apparently, there was no discussion about safety issues associated with the alteration in service parameters. There was no documentation exchanged between either party attesting to the contract alteration and the lowering of service levels.

On Dec. 23, it snows lightly and the north side of the street in the HOA receives sunshine, while the south side remains in shadow for most of the day. Within a few hours the snow on the north-side driveways and walks are bare because of the sun exposure, However, the south-side driveways and walks turn icy over the course of the day.

That evening one of the homeowners hosts a Christmas party and during the party several guests mention the icy drives and walks along the south side of the street. On the way home from the party, one homeowner slips and hits his head on the drive. A lawsuit ensues. ABC is named in the suit, as is the management company and the HOA.

During depositions it is discovered that ABC Snow was aware of the tendency for ice to form on the south-side drives and walks. In previous visits, ABC laborers paid attention to that side of the street so as to keep the site safe. They kept accurate records of site visits, services rendered, materials used and quantities of product applied.

ABC has been in business more than 10 years, is a member in good standing in various trade organizations and the management is knowledgeable about the latest snow and ice removal techniques and the importance of accurate record keeping. When called about the accident, they immediately dispatched personnel to apply deicing material to the homeowner's driveway and walks. Subsequent examination of the service records indicates the company serviced other sites in the area during the day of the incident, but did not service the HOA in question.

Deposition testimony reveals the property manager assigned to this HOA claimed to have an unclear memory of the alleged telephone conversation with an ABC representative that reduced the service levels for the HOA. In fact, the property manager doesn't recall the conversation, or what might have taken place during this alleged conversation.

ABC is a good snow contractor. They have always proven to be contentious, educated about the industry and professional in their actions in the field. However, in the opinion of the expert witness, this contractor was smart enough to know documenting such a drastic and dramatic alteration of the site's service contract definitely required written documentation to protect the company's interests.

The plaintiff's position was that the contractor should have documented such a change in the scope of work. In the absence of such documentation, and considering the property manager failed to recall the conversation, then the snow contractor was at fault.

Correspondences between the HOA, the property management company and the snow contractor would have placed responsibility on the appropriate party. Even in the absence of such documentation, had the property manager recalled the conversation and admitted to lowering the site's service levels, it is likely the property management firm would have been responsible for the situation that resulted in the accident.

Further, had the telephone conversation with the snow contractor occurred, the property management company should have documented the scope of the contract alteration to the HOA's board attesting to the fact the HOA's board directed the change. Of course, no one from the HOA's board recalled directing the property manager to alter the scope of work due to budgetary constraints.

The plaintiff's expert witness opined the contractor was too smart to have just "forgotten" to document the discussion via correspondence with the property management company. It was not unusual for the property manager to be forgetful of any such conversation. Nor, was it unexpected the HOA's board members would conveniently forget about any directive to the property management company to cut back services due to lack of funds. The HOA had no documentation from anyone attesting to the change in the scope of work required.

RESULT

As with most cases, this one settled. The plaintiff was happy with the settlement, although all parties had to consent to a non-disclosure clause as part of the settlement agreement. Undoubtedly, the insurance companies for the defendants paid out some monetary consideration to the plaintiff. Regardless of the amount, the snow contractor's insurance company will have taken some sort of "hit" on its reserve for this incident. The snow contractor lost a (previously) good client. The plaintiff suffered long-term disability that no amount of money can correct. There is no doubt the snow contractor spent copious amounts of time defending this suit – time better focused on productive activities.

LESSON LEARNED

Everyone was left holding the bag and the snow contractor was at the bottom of the totem pole with regards to responsibility exposure. A simple email or letter to the property management company documenting the conversation and the alteration in the contract would have easily shifted all responsibility away from ABC Snow Contracting and back to the property management company and/or the HOA. When in doubt, document it.

CASE STUDY 2:
ONE SLIP AND YOU CAN TAKE THE FALL

BACKGROUND: At an East Coast retirement facility – we will call it The Center – a long-time tenant exited a building and walked to his car. The configuration of the sidewalks and parking lots was such that cars are parked immediately adjacent to the walks. To access their vehicles, pedestrians needed to step over snow piled on the curb line. There were no provisions for "curb cuts" that would have allowed unfettered access to the parking areas.

While attempting to get over the snow "hump" on the curb, the tenant slipped and fell on a patch of ice just off the curb where snow was piled. After the fall, several tenants gave deposition testimonies that there had been ongoing issues with regard to whether the paved areas had been maintained adequately prior to the incident.

Many who worked in the building or for The Center indicated inadequate attention was given to proper deicing. Deposition testimony revealed the employees of DFG Lawn Care (not the company's real name), which was responsible for winter exterior maintenance, did not received formal snow-and-ice-management training. DFG's owner admitted he did nothing to further his own professional training in generally accepted industry practices for snow and ice removal, and he was basically "self taught" through on-the-job experience.

Testimony indicated few, if any, return visits were made to the site after the initial clearing of snow, even though the contract documents specifically called for a mid-day return visit to perform "clean-up" work and additional deicing. It became evident that return visits to the site would have allowed the company to clear the snow and ice that had accumulated between parked cars after the initial plowing. It is important to note that the tenants would periodically relocate their cars in the anticipation that further service would rid those areas of the snow and ice not addressed in the initial visit.

Weather data supplied by an independent weather expert for the date in question, as well as five weeks prior to the accident, indicated

conditions in the general geographic area were conducive to having slippery and icy conditions present at the time of the accident. The weather records indicated there was snow on untreated ground surfaces up to and including the day of the accident. With the weather records indicating temperature swings, it became apparent the area experienced several thaw-and-refreeze cycles in the weeks and days leading up to the accident.

SNOW FACT:
New York City used more than 300,000 tons of salt during the 2010-2011 winter season for deicing.
Source: New York State Department of Transportation

Even though the contract provided guidance for return visits to perform additional snow clearing and/or deicing services without prior approval or notification from the client, there was no evidence of any activity on the part of DFG Lawn Care in the seven days prior to the accident. If you consider the weather data for the area, it was inconceivable that no site visits were made to check the conditions during this period. The contractor was not under any restrictions that would prevent such activity, and was, in fact, tasked with doing just that on an as-needed basis.

The absence of any evidence that the contractor was prevented from performing a site visit was justification enough that such monitoring activity was acceptable. The wife of DFG Lawn Care's owner specifically stated that if monitoring services were provided at this site (monitoring for obvious thaw and refreeze activity is a service provided by snow professionals in this and other markets), the customer would be charged. The manager of the site specifically stated The Center never questioned the invoices from DFG Lawn Care – thus, such activity would have been chargeable and accepted by The Center as valid.

There was absolutely no reason for DFG Lawn Care not to check the site under these circumstances. In fact, there were many reasons – given the evidence provided – that checking the site should have warranted the application of deicing product and performing deicing services:

- Weather data supported thaw-and-refreeze activity.
- This site had a general slope that encouraged moisture run-off facilitating refreezing when the temperatures dropped at night.
- The contractor had no restrictions that would prevent him from inspecting or providing deicing services.
- The contractor would clear individual parking spaces, but acknowledged no return visit would take place to clear parking spaces vacated after the initial plowing. This was done even though doing so would have eliminated most, if not all, of the snow that accumulated between, in front of, and behind the cars. This would have eliminat-

To avoid costly mistakes, pay close attention to contract details.

Surfaces that pedestrians frequent require careful attention and monitoring by contractors for ice and snow buildup.

ed unsafe thaw-and-refreeze conditions in those areas.

- The contractor was paid "by the hour" for plowing, and "by the bag" and "by the ton" for deicing. Deicing can be a profitable portion of a contractor's business, thus negating any argument against providing the service, especially in light of the contract, which specifically stated it was DFG Lawn Care's responsibility to do so.
- Generally accepted practices in the snow industry call for monitoring and deicing applications given the circumstances indicated by the deposition testimony, the contract documents and the evidence supplied.

Snow industry professionals throughout the United States normally pursue ongoing training for owners and employees. Membership in a professional organization dedicated to the landscape and snow and ice management industries that provides ongoing training would be evidence of professionalism.

DFG Lawn Care presented no evidence of even rudimentary training of its staff. In fact, DFG Lawn Care employees testified they received absolutely no training on equipment operation, service performance, recognizing the need for service on any site or any training in the snow industry.

One tenant testified during his deposition that the sidewalk snow was shoveled onto the narrow 12-inch-wide grass area between the sidewalk and the parking lot. This created an artificial accumulation of snow, which continued to melt and refreeze onto the parking lot. Further in his deposition, the tenant indicated DFG Lawn Care plowed straight down the middle of the parking lot with blades angled toward the rear of the parked cars. Residual snow would come off of the plow and go to the right and left of the lot under and between the parked cars.

DFG Lawn Care did not shovel or salt between the cars – they were not contractually obligation to do this – and DFG did not request that tenants move their cars. The tenants were notified DFG Lawn Care would return during the noon hour to perform clean-up work, including plowing and removing snow from the curb lines and between parked cars that were now relocated. The failure to return, as contractually obligated, created an artificial accumulation of snow between the cars that continued to melt and freeze, and caused the hazardous condition that contributed to the accident.

The plaintiff indicated The Center was also responsible for this incident. Their employees others visiting the premises. It was alleged that when there is melting and refreezing, The Center would have a duty to call DFG Lawn Care to salt the parking lot. Further, The Center was aware that there are elderly and infirm tenants in this building, and that there is no alternate route for the tenants or visitors to the facility to use to enter the building.

> "The contractor would have had a stronger legal leg to stand on if they had **followed the contract** language and performed service accordingly."

testified they had no responsibility for oversight on the exterior areas of the building or the parking lot. From time to time, they might have walked the parking lot looking for signs of icy conditions *if* enough complaints were received. This lack of caring for pedestrian, tenant and vehicle-operator safety showed a complete lack of adherence to even the most basic site-management principles.

It was alleged The Center would have an independent duty to protect its tenants and The plaintiff's expert witness testified the contractor had not followed the language in the contract, had not educated himself (or his employees) of the local standard of care for that region and generally ignored all accepted protocol for serving the site properly. Therefore, it was his professional opinion that the situation that occurred could have been avoided had the site employees and/or the snow contractor instituted and practiced generally accepted principles of snow and ice management on this site.

RESULT

The case was settled and both DFG Lawn Care and The Center's insurance carriers paid out significant sums of money to settle the claim. It is likely the insurance carrier for the snow contractor admonished them for failing to follow the guidelines of the contract documents. Additionally, The Center's management changed snow contractors and asked considerably more questions about contractors' qualifications before hiring them.

LESSON LEARNED

The property owner learned to thoroughly research outside contractors to determine whether they were hiring a true professional. Had the contractor read and understood the contract language, and had they been better educated on industry practices, it is likely the entire incident could have been avoided. The contractor would have had a stronger legal leg to stand on if they had followed the contract language and performed service accordingly.

CASE STUDY 3:
A WEIGHTY, COSTLY ISSUE

BACKGROUND: The snow contractor, we will call them BEST Snow Removal, had the contract to clear snow from a retail facility that included a multi-level parking ramp. BEST Snow Removal subcontracted the actual on-site work. However, the subcontractor was not well versed in how to address large snowfalls on the top deck of a parking ramp.

This was in a market that received an average snowfall of 35 inches each season. Following an unusual 20-inch snow event, the subcontractor attacked the snow accumulation on the deck with an F-450 truck and a 30,000 lb. loader. Snow clearing operations began about the time the snowfall ended. During the clearing process, all the snow was piled at one end of the deck.

The resulting snow piles were estimated at 12 to 15 feet and extended out 30 feet from the edge of the deck. The snow was not actually removed from the deck, nor was the snow lifted over the end walls and deposited on the ground. Within 36 hours of the end of the snow clearing operation, a large portion of the upper parking deck collapsed, pancaking down through the two deck levels, settling on the ground level. Fortunately, no one was injured as the collapse occurred during overnight hours when the retail facility was closed. The facility owner filed suit against the snow contractor who signed the original agreement and the snow contractor enjoined the subcontractor, which is normal under the circumstances.

An investigation of the engineering surrounding the construction of the structure showed the parking garage was designed to accept a total load of 63 lbs. per square foot. This figure reflects 40 lbs. per square foot for vehicle weight, 21 lbs. per square foot for snow loading, and 2 lbs. per square foot for miscellaneous dead loads.

The weight of snow varies greatly. Light fluffy snow may only weigh about 7 lbs. per cubic foot and average snow may weigh 15 lbs. per cubic foot while drifted, compacted snow may weigh 20 lbs. or more per cubic foot. For comparison, consider these weight-related facts:

- 7.48 gallons of water in 1 cubic foot.
- 1 cubic foot of water weighs 62 lbs.
- 1 cubic foot of ice weighs 57 lbs. (essentially 1.5 – 1 snow)
- 1 cubic foot of 5-1 snow weighs 25 lbs.
- 1 cubic foot of 10-1 snow (cotton ball snow) weighs 10.4 lbs.

The snow from the plowing operations was piled on the parking deck surface and not lifted over the side. The expert witness expressed that it was his professional opinion, based upon the facts submitted in the case, that the removal methods used at this site were incorrect. Generally speaking, most parking deck snowplowing is done with the specific intent of removing excessive snow from the deck area while plowing operations are in progress, or immediately thereafter.

In this case, given the amount of snow that ultimately fell at the site, snow removal operations should have started at the time plowing operations kicked off. One can argue that waiting until the snow fall had stopped to begin plowing was prudent. But given the enormity and length of the storm, it was an error in judgment. One might also argue that most snowstorms don't leave this much snow in their wake. This storm was predicted days in advance and snowfall totals were projected in excess of what actually fell.

The use of an F-450 plow truck with a loaded salt spreader in its bed was indicative of the lack of knowledge and experience on the part of the operator. Incredibly, a 30,000 lb. articulating loader was also used on the deck. Not only is it common industry knowledge that the largest loader unit that should be used on a parking deck is a skid steer under 5,000 lbs. gross vehicle weight, the use of a 30,000 lb. loader on a parking deck flies in the face of any common logic.

The pictures of the site after the deck failure clearly showed damage to the exterior wall of the deck structure. This shows the operator was attempting to use the wall as a backstop to push against. The weight of the loader alone far exceeded the weight limits recommended for this structure and when you factor in the weight of the snow that was intentionally piled up on the deck, the likelihood of structural failure was a forgone conclusion.

It was determined the operator(s) and owners of the plowing equipment acted in-appropriately and unprofessionally in using oversized equipment on this site for plowing accumulated snow. It was determined that waiting until *after* the snowfall stopped was inappropriate given the advance knowledge of the type of snow event that was to occur, and piling snow to such a height on a parking deck fell outside accepted industry norms. All these

SNOW FACT:

THE SNOWBELT
Traditionally, the Northeast and Midwest playhost to the most severe snowstorms. The area defined as the Snowbelt, stretching across the Great Lakes from Minnesota to Maine, usually receives the brunt of winter storms.

factors individually would have likely led to the structure's failure, but when combined the chances increased. It was also determined the contractor who signed the original agreement should have better educated his subcontractor on how to best address the situation.

RESULT

Given the overwhelming evidence, the case was settled quickly. BEST Snow Removal had to participate in the settlement and its insurance carrier had to pay out. The subcontractor bared the brunt of the settlement because he represented himself to be a professional snow contractor, having been in business for well over a decade. Thus, he should have been experienced enough to realize taking a loader onto the upper deck of a parking ramp was not the right call. To be sure, everyone learned valuable lessons about addressing such situations.

LESSON LEARNED

The property owner learned to include language in agreements about structural integrity with regard to snow accumulations on parking decks. The contractors learned what can happen if too much weight is left on a deck. Both parties learned, albeit the hard way, two new words – snow loads.

Chapter Seventeen
STANDARDS AND PRACTICES

Chapter Highlights

- ○ Zero Tolerance
- ○ Thaw and Refreeze
- ○ Return Visits

As the industry has evolved and matured, outer-ring contractors – landscape, tree care, irrigation, excavation – have become better educated and more sophisticated. At the completion of this edition of "Managing Snow & Ice," written standards detailing how snow contractors should service sites do not exist.

It's a tall order. Standards, as we now know them, often stem from the customer's tolerance levels – and even these vary. For example, the standards that shape the industry along the East Coast are very different than those in the lake-effect areas of the Great Lakes region. Overall, the snow industry defines "tolerance" as the amount of snow on pavement that is acceptable and

considered nominally safe for vehicular and pedestrian traffic. In simple terms, it is the level of service expected, or demanded, by customers.

I have had the privilege to provide snow services in virtually every geographic region in the United States and Canada. I have also visited and worked in 36 countries in Europe and Asia, and spent time working in the Arctic and Iceland. This experience allows me to determine the tolerance levels for a good portion of the world that experiences snow.

Zero Tolerance

"Zero tolerance" is wet or bare pavement at all times. While not confined exclusively to the East Coast of the

Certain accounts, such as medical facilities, require "zero tolerance" for their pavement.

173

Residential customers expect their driveways to be cleared in a timely fashion.

United States, this service level is most prevalent in markets commonly referred to as "the I-95 corridor" – from Boston to Washington D.C. This level of service is requested more often in retail situations where customers experience higher levels of pedestrian and vehicular traffic. Thus, the potential for slip-and-fall accidents is greater than at industrial and other commercial sites.

To achieve zero-tolerance levels, a site must be plowed and treated on a constant basis. For continuous snowfalls, repeated plowings are necessary to remove additional snow accumulation. Likewise, additional applications of deicing material are required to mitigate ice buildup. For less than 1-inch-per-hour snowfalls, continuous bare and wet pavement is achievable as long as continuous attention is paid to the site.

For snowfall rates of 1.5 inches or more, bare and wet pavement is more difficult to maintain. At a snowfall rate of 2 inches or more per hour, even the best contractor will be hard pressed to keep a bare-and-wet site at all times.

As a rule, condominium or home owner's association (HOA) sites require a higher level of service than industrial sites. However, tenants or owners of these associations are sometimes reluctant to pay for such service due to extremely tight budgets, lack of knowledge on the part of the tenants or owners and a propensity for thinking they know more about the snow industry than industry professionals. Providing a bare-and-wet service is normally not the standard for these sites.

Industrial facilities don't have the continual, high-traffic scenarios found in retail settings, and thus normally demand a lower level of service when it comes to snow and ice on their parking lots, drive lanes and sidewalks. Contractors often refer to this as the get-them-into-work philosophy where the expectation is to have the site plowed and deiced so workers can get into work. After that point, though, the service expectation level is lowered.

Office parks that have multiple tenants normally have higher service level expectations than industrial settings, but lower than retail settings. Workers require access to the buildings in the morning and the ability to depart the property in the afternoon. Overnight and mid-day plowings and deicing treatments are not as important as making sure the site is cleared at the beginning and end of the work day.

Customers demanding a 1-inch trigger for the start of services are willing to accept some of the liability exposure for snowfalls under that amount. "Bare and wet" is not achievable with trigger depths for snow clearing operations that begin at 1 inch or above. Some snow will be left on the pavement surface. Even with regular applications of deicing material, some snow usually remains.

When sites are cleared of snow and deicing material applied in the morning hours, return visits that evening to clean up the slush that results from salt and vehicular traffic are normal. This, however, is not as common sense as some would be led to believe. This type of buildup can be dangerous to vehicles and pedestrians if not cleared off the pavement surface. The fine layer of ice that can build up must be treated with deicing material to reduce the ice and snow to water.

The public's tolerance for snow and ice on pavement and sidewalk surfaces varies by geographic market. In the Northeast and mid-Atlantic, the general public demands zero-tolerance conditions for retail sites, especially parking lots. Sites dominated by office buildings that demand zero-tolerance want greater safety for both pedestrians and vehicles.

The Midwest is becoming more attuned to safety issues with more and more customers demanding zero tolerance conditions. One-inch triggers are common place in these markets. Go further west and those markets lean toward a 2-inch trigger depth. In contrast, mountainous areas mostly tolerate snow on all pavement surfaces. Having some snow is normally preferred to wet-and-bare levels in these markets. It is difficult to achieve wet-and-bare conditions when temperatures are well below 32 degrees F for extended periods of time. It takes a very long time for rock salt to work in temperatures below 15 degrees F.

Snowbelt areas south and east of the Great Lakes tolerate snow more than any other geographic area of the country. In some areas within this region, plowing operations do not even begin before 3 inches of snow has accumulated on pavement surfaces. Streets with 4 to 6 inches of snow are regularly left to go without any plowing operations at all.

It is the rare customer who allows snow to be piled up against the building. The expectation is that the snow should be removed and plowed away from the building's entry and exits.

Snow should be piled in green spaces and/or in designated snow-dump areas where thaw-and-refreeze cycles will not endanger pedestrian and vehicular traffic. Generally, these areas are in the outer regions of the parking lot or on the side of a roadway furthest from the building.

Snow should never be placed in handicapped parking. Certain states, such as New Jersey, have laws specifically prohibiting snow from remaining in handicapped parking.

Avoid placing snow on landscaped islands, which often run parallel with the building and just outside the drive lane. This can damage plant material and would require replacement in the spring. Likewise, water run-off from snow piles can create dangerous walking conditions due to thaw-and-refreeze cycles.

Snow removal can be a 24-hour business when a major snow event is in progress.

Thaw and Refreeze

Thaw and refreeze occurs when snow begins to melt. Remember, snow does not melt from the top of the pile. Snow packs down over time even if air temperatures are below freezing. In fact, some snowmelt occurs with temperatures at, or slightly below, freezing.

Water discharge from the bottom of the pile runs towards the lowest point on the pavement. At night, this run-off will likely refreeze if air temperatures drop below the freezing mark. This newly formed ice must be removed. Normally, the contractor uses an appropriate deicing product to either melt the ice or reduce the freezing point to allow the ice to return to a liquid state.

Generally speaking, placing snow in spots that promote run-off across the pavement is frowned upon. Sometimes the configuration of a given parking lot requires that snow be placed at a high spot on the property. Particular care needs to be taken to protect the safety of those using a paved surface where run-off is unavoidable. Unfortunately, some customers do not understand what happens in thaw-and-refreeze situations, and may dissuade the contractor from returning to address these scenarios.

It is wise for contractors to protect their companies by writing down any conversations or directives of this nature they have with customers. Verbal interactions can often be misunderstood or not remembered accurately. With today's communication technology, sending an email to confirm details and instructions protects all parties.

Return Visits

A return visit to a site after the storm has ended seems like common sense. When customers are being charged per event, contractors generally write into their agreements that one return clean-up visit is part of the per-event price. However, for per-push contracts return clean-up visits result in additional charges to customers.

Some customers don't want to pay for this type of service. However, these return visits are necessary to provide a safe environment for employees and visitors to the site. Sometimes additional plowing operations are unnecessary, but an additional application of deicing product is required to keep the site safe. If customers do not want return visits, the contractor should obtain a written statement from the customer stating no return visits are to be made after completion of the initial clearing and deicing.

Contractors who include language in their service agreements that says they are not responsible for anything that happens on the site are kidding themselves. That's no different than putting a sign in the rear window of your car stating, "I'm not responsible for accidents I might cause."

Today's snow contractors are risk managers, and it is their jobs to keep sites safe. Customers who do not wish to pay for this service need to be advised in writing of the potential harm resulting from a failure to keep a site safe both during and after a snow event.

NOTES:

Chapter Eighteen
THE FUTURE OF SNOW AND ICE MANAGEMENT

Chapter Highlights

○ A Look At The Future Of Snow and Ice Management

What does the future hold for how contractors manage snow events? This is difficult to predict. When this book's first edition was written, Nextel's Direct Connect only served small regional markets. You could not Direct Connect across state lines or even from Philadelphia to Pittsburgh. Cell phone calls from another geographic area added roaming charges to your bill. In fact, plowing "outside your market" meant going 50 miles away.

Texting was not a popular communication tool. Laptop computers were heavy and cumbersome compared to today's ultra-light models. The Weather Channel was mainly a television-only experience, and looking at weather radar was limited to the desktop computers and laptops in your office.

Contractors communicated mostly via two-way radios and CB radios. We needed to consult a physical map for directions to go anywhere. Likewise, we'd visit each site to measure it, and then rely on experience to determine how long it would take to do the necessary work.

Today, cell phones and mobile devices allow a contractor to email customers in seconds from the office or in the field at a job site. Reporting on the progress of a job from the field can be done by calling into a computer. That computer, in turn, provides details on the account and where you need to go next. Very soon, an "app" for the iPhone will streamline reporting from the truck or salter in seconds.

The iPad and other tablet-style devices allow access to the Internet from just about anywhere – the truck, the garage, Starbucks or McDonald's. Contractors can start trucks from 100 feet away with remote keys, and GPS

> "Very soon, an "app" for the **iPhone** will streamline reporting from the truck or salter in seconds."

technology allows you to go anywhere and never get lost, as long as you have an address or the name of the facility that needs service.

A contractor can provide a quote for a site and never actually leave his office. You can view sites from above, the side and the rear. You can virtually walk around the building from the office and see every detail of the site. You can measure the site, count the plants and trees, determine the pitch of the parking lot, figure out how tall the building is and learn who the neighbors are with the click of a mouse. You never have to leave your desk.

Technology advances will play an important role in the development of your company.

Estimating packages use verified production factors for all sorts of combinations of equipment to determine how long it takes to clear snow and deice a site. A snow contractor can accurately quote an entire city without ever actually visiting any of the sites. Even with the technology available to us today, nothing really can replace getting a in-person look at the site during the process. I would never just send the troops to a new account without having laid my eyes on it. Using technology for quoting purposes is fine, but a site visit is necessary to develop a detailed attack plan.

Back in 2002, no one dreamed of the great technological advances that would impact our industry. Blizzard plows had not yet been invented and only one plow manufacturer had a viable "V" blade. Pushers had only been on the market for a few years. In fact, whole regions of the country had no idea what a snowpusher was, how it worked or what it could accomplish. At that time, salt spreaders were all made of steel.

The times and the technology have changed dramatically, but contractors still have plow and provide exceptional customer service. This hasn't changed. It does, however, make one wonder what the industry will look like in 2020?

A Look At The Future Of Snow and Ice Management

Making predictions can get you into trouble. The proverbial crystal-ball method of trying to guess which direction the industry is headed involves some considerable guesswork. I wonder if I would have predicted some of the advances that came about over the past 10 years when authoring this book's first edition. But what the heck, here goes nothing:

No. 1. The industry will continue to evolve and technology will make our lives easier from an administrative standpoint. In fact, I envision enhanced reporting scenarios that will allow complete interaction between what the crew is doing at the site, billing of the customer and billing from the subcontractor to the plowing contractor. This would all interact with any standard accounting software available on the market. To achieve "automatic" reporting today, though, requires an operator equipped with interactive voice recognition (IVR). But wouldn't it be nice if the truck did the reporting through a modified GPS device? It would eliminate from of the equation all potential for reporting error. The truck "reports" when it arrives at the site, what is done and when the work is completed.

No. 2. How about having a digital camera that records video of the work being done at the site. The contractor rotates the camera on a swivel to capture the work that has been done. This video evidence would protect the company from bogus insurance claims. You could also provide select customers with access to the real-time feed so they can see what you're doing at their sites.

No. 3. Computerized sensors embedded in spreaders will automatically calculate how much material must be applied to the pavement. This same device also calibrates the spreader/spinner to put down the correct amount of product needed for the conditions present at the site. It then changes automatically when the truck arrives at the next site. Plus, it sends information back to the office for billing and expense calculations.

No. 4. Other than what is mentioned in this book, there are no written industry standards. The hope is there will be a set of comprehensive and detailed written standards for North American snow contractors. In addition to guiding snow professionals, these standards would allow the insurance industry and legal community to better protect the public by demanding contractors abide by these standards.

No. 5. I envision an established college or university developing curriculum for educating front-line workers, middle managers and upper management on the procedures necessary to provide quality customer service. It would be a full-blown program with candidates graduating with a certificate or degree in snow and ice management. Current certification programs measure knowledge and encourage business owners to become educated in business issues, but a formalized curriculum would teach individuals how to do the jobs required on the site and in the office.

No. 6. The current business model for the "nationals" is not sustainable. I believe they will be replaced with a better, more feasible model. This new model will be based on providing quality service at fair pricing, instead of promising lower prices that come from forcing contractors into substandard work to secure the account.

No. 7. I see a nationwide "roll-up" of independent snow contractors. This will form a national marketing or buying group. This entity would be backed by venture capital or angel financing, and it would bring a dozen or so regional providers to the table. As noted earlier in No. 6, I do not believe the current national service providers and their business model will survive.

No. 8 Over the next decade, will see our first publicly traded snow management company. It is inevitable. Some enterprising person will obtain the right pricing and do the aforementioned roll-up strategy with the specific intent of taking the company public. It will struggle under its own weight, and it may eventually end up being acquired by some larger entity. However, I believe the industry will be ripe for this type of entity to exist.

SNOW FACT:

Each year an average of 105 snow-producing storms affect the continental United States. A typical storm will have a snow-producing lifetime of two to five days and will bring snow to portions of several states.

Chapter Nineteen
MANAGING WEATHER RISK

Chapter Highlights

- Per-Event/Push Risk
- All-Inclusive Risk
- Mix Of Contracts
- Things To Think About
- How To Get Started

The primary risk a snow management contractor has is being at the mercy of Mother Nature – is there too little or too much snowfall. Each company's risk is different based on the contracts they have with clients and, of course, their geographic location.

How can snow contractors minimize this risk – aside from carrying a rabbit's foot or horseshoe in their truck all winter? They can consider the services provided by Chicago Weather Brokerage, which helps to protect snow contractors and their customers against adverse winter weather conditions, whether that be low-snowfall or excessive snowfall.

Protection levels are available on a monthly and/or seasonal basis, and contractors can establish positions best representing their risk levels based on their given geographic region(s). This new business model can help contractors better predict revenue flow and profitability levels.

The following is a review of how Chicago Weather Brokerage applies this financial hedging tool to various types of plowing contracts in the industry.

Per-Event/Push Risk

Contractors do not control when Mother Nature delivers revenue generating snow events. There can be several significant snow events one month and none the next. To determine your company's snowfall risk, it is recommended you look at the numbers one month at a time.

Review each month in the season – November through March – and identify the average snowfall. It is best to look at a 10 year average and determine your revenues in an *average* year. For example, the average December snowfall is 10 inches. Your average revenue at 10 inches of snow is $50,000 or $5,000 per-inch. Quantifying your revenue per inch of monthly or seasonal snowfall is a key component to successfully managing your business' weather risk.

Your next step is to determine what happens below the average. For example, at 5 inches, your revenues would average $25,000. Then look at your monthly expenses – what do you need to cover your bills or to cover your minimum desired profitability? Your hedge position will be structured in a way to cover these desired minimums

The reason you measure on a monthly basis is because that is how payroll and your expenses are incurred and using this strategy allows you to never have an issue with either. You will need to go through this process for each month of the season.

The snow industry is unpredictable and at the mercy of Mother Nature.

All-Inclusive Risk

This risk is generally a seasonal risk but can be done for the monthly model as well. It is not often that Mother Nature delivers average snowfall for a season in one just month and if she does, and a company did not protect itself, it can always hedge later months during the season, but not as effectively.

To determine your seasonal risk for all-inclusive contracts, use the 10 year average snowfall. At that average snowfall, determine your profits on your all-inclusive business – it should be close to 50% – and then determine the per-inch profit.

For example, if the seasonal average is 40 inches on an account that generates $100,000 in revenue and $50,000 in profits, at the average, the profit per inch is $1,250. Then look at the risk. At 5 inches above the average, you would lose approximately $6,250 in profit and 10 inches above average you would lose $12,500 in profit.

Mix of Contracts

The most common mix of contracts for a snow removal company is 70% per-event and 30% all-inclusive. This is a hedging strategy just like the financial markets provide but snow and ice management

contractors still have significant risk.

A company can certainly pay its bills with the all-inclusive contracts, and per-event contracts cover a lot of the all-inclusive losses in extremely high-snow-volume winters. However, the likelihood of the company "taking a hit" on one or the other during the year is quite real but this risk can be managed.

There is a significant advantage for the company if its contracts are set up this way. Your company's risk is only at the very extremes of snowfall and it is more cost-effective to establish a hedge position. You need to go through the processes above and determine your risk for *both* types of contracts by the inch.

For example, if there are 40 inches of snow during the season broken down this way: November (4 inches); December (6 inches); January (14 inches); February (10 inches); and March (6 inches).

There is a minimum level of income each month from the all-inclusive contracts. This usually means that your company is not concerned about anything close to the average low snowfall. However, the concern is the extreme snowfall. In this scenario, the December average is 4 inches and if your geographic area has 1 or 2 inches of snow, you are going to cover cost but have virtually no other income. Thus, 2 inches or less would be your hedge position.

For all-inclusive contracts, it doesn't matter how much your company is making on your per-event business. No one wants to service an all-inclusive account when you are double the average snowfall and losing money.

Using the previous example, at 80 inches of snow you have eaten up *all* of your profits on that account - $1,250 x 40 inches = $50,000. You need to determine the level that you want to establish a hedge position and with a per-event revenue model, it will most likely be close to 70 inches. Yes, this is extreme, but that is what the process is all about, eliminating the extremes from the business. It's not about how much you make, it's about how much you risk relative to how much you make.

As you review the information outlined in this chapter, let's look at a few examples of average snowfall totals in major snow markets:

Denver, December 2010: Average snowfall is 25 inches for Denver; actual snowfall was less than 5 inches of snow.

Chicago, winter of 2010-2011: Average snowfall is 36.94 inches; as of March 2011 actual snowfall is at 56.3 inches.

New York, December 2010: Average snowfall 7.93 inches; December 2010 actual snowfall was 13.7 inches.

Des Moines, December 2010: Average snowfall is 13.7 inches; December 2010 actual snowfall is .08 inches.

The snow management industry is a business of extremes and this service can allow your company to manage them, and most importantly protect its revenue and profits. However, before you decide here are some things you need to think about and tips to get started.

Things To Think About

- What is the worst case scenario for my business?
- If that scenario occurs, what are implications to my business?
- What is preventing the growth of my business?
- Where do I have risk in getting paid?
- If there are negative financial implications based on what Mother Nature does, then you are ready for this product.

How To Get Started

- Determine the risk you need to protect against (based on inches).
- Calculate the amount of risk in dollars.
- Contact Chicago Weather Brokerage and talk through that information with them. They will tell you how much it will cost manage that risk.
- You will need to open an account with a clearing house. Chicago Weather Brokerage will let know which one.
- You will need to deposit your money in the account at the time of the transaction.
- This process takes some time, so plan early.

SNOW FACT: Several less-populated areas around the country that receive a lot of snow. For example, Mount Washington, N.H., has an average annual snowfall of 260 inches, and Valdez, Alaska, averages 326 inches annually.

FOR MORE INFORMATION
Learn more on the Chicago Weather Brokerage, at www.cwbrokerage.com

Epilogue

John Allin's career managing snow and ice stretches back to his childhood when his father – a welder by trade – plowed snow as a side business. From that beginning, Allin has traveled the globe teaching, consulting and talking about snow and ice management. Along the way, he has built and sold several businesses, presented lectures to thousands and helped develop new equipment technology. Today, he shares his knowledge with those seeking to learn how to perfect the art of managing snow and ice. Here are some thoughts and insights from Allin as he reflects on his career and what it has meant.

In The Beginning

Back in 1990, I attended a trade show on snowplowing in Lansing, Mich., and it was there I met Kyle Hansen. We hit it off immediately and became close friends. At the time, I was running about 35 subcontractors in the snowplowing portion of my business in Erie, Pa., and he had recently taken over running his father's power-sweeping business in Minneapolis, which had more than 150 vehicles moving snow.

One of the sessions at the show dealt with how snow contractors priced their services. One person insisted that all pricing – regardless of location – had to be per-season pricing and there was no other way to do it. As the discussion heated up, I asked an innocent question, "How many snow events do you base your 'seasonal' total on?" He answered, "Six." I looked at the fellow next to him and posed the same question and he replied, "Five."

I found this interesting and asked how much snow they received in their respective markets. The answers around the table were just as puzzling. One said, "46 inches in Madison, Wis.," "54 inches in the Hudson Valley area of New York," and another said, "38 inches in the Minneapolis/St. Paul area."

When it was my turn, my reply to the same question was such that the entire audience was dumbfounded. "In Erie, we base our seasonal pricing on 36 plows, and we get in excess of 250 inches of snow about three miles south of the immediate lakeshore." I prefaced the statement by saying, "You're all going to think I'm nuts," and their reactions reflected as much.

The interesting thing to me was that I thought everyone in the United States got snow like we did in Erie. I believed that hundreds of inches of snow were normal and not the slightest bit unique. During my call home to my wife, Peggy, I told her how amazed I was to discover that people had viable snowplowing operations with anywhere from 30 to 60 inches of snow each year.

The next day this same group went to the trade show and we stopped by a booth that was selling weather reporting services. Remember, this was in weather reporting's infancy for plowing contractors, and we proceeded to question the representative extensively.

A very accommodating individual, he answered us as best as he could. He explained the company had reporting stations across the northern portion of the United States – one in Madison, Wis., the Chicago area, and St. Paul, Min. When I said I was from Erie, he looked at me long and hard and stated, "We can't predict the snow in Erie." He went on to say that, due to the lake effect, a two-degree shift in the wind direction off of Lake Erie could mean the difference between getting 2 inches or 2 feet of snow. I was gaining credibility with my peers as the company rep corroborated my story.

At a gathering of plowers at dinner that evening, Kyle Hansen came up to me and said he wanted to talk. He was having trouble dealing with heavy snowfall totals that occasionally hit his Minneapolis market. I told him I really wanted to talk to him as I was having considerable difficulty moving past the 35-unit mark and wanted to expand.

Interestingly enough, Kyle began ticking off all the issues we were facing in Erie in our quest for growth. And I was able to do the same for him with the deep snowfalls he experienced in Minnesota. This discussion took place during a long walk from the restaurant back to the hotel. During that walk we developed a close friendship that continued for well over a decade.

Following the show, I was asked to join the snowplowers association's board of directors. Shortly thereafter, Kyle Hansen and Marc Young were also asked to join the board. We were honored to do so and believed the organization was sorely needed, as the industry needed to achieve a greater degree of professionalism.

A year later, Minneapolis hosted the show and I was asked to speak about how we dealt with deep snowfalls in the Erie market. I took on the challenge and thus began my speaking career about snow and ice issues. Following the meeting, Marc Young, Kyle Hansen and I were approached to take over the operation of the association.

After lengthy negotiations, we offered to purchase the rights to the name and the membership list for the sum of $1 along with a promissory note in the amount of $10,000 – with no set terms for repayment.

Our plans were to make the organization non-profit and grow membership beyond the current levels. Our offer was politely refused and the association was allowed to fold. Hindsight proved to us that this was for the best since the association did not have a good reputation with many industry professionals.

A Seed Is Planted

Several years passed. In January 1996, I gathered the business cards I had obtained from my travels speaking and sent a letter to each one asking if there was any interest in starting a new, non-profit association dedicated to advancing the cause of the professional snow industry. We received 23 responses and, of course, Kyle Hansen and Marc Young were among the first to respond in the affirmative.

In April 1996, we mailed another letter to the 23 respondents and told them if they were serious and wanted to be part of creating a non-profit association, then they should be in Erie on Friday, June 6, at Sid's Restaurant at 7 p.m. for dinner. In addition, they were required to make the trip at their own expense, and had to be prepared to spend the entire weekend working out the details.

Thirteen individuals responded and nine actually showed up. Those original pioneers were Kyle Hansen, Marc Young, Phil Christian, Rich Redfern, Charles Glossop, Bill Milbier, Rick Kier and Jeff Tovar.

Dinner that Friday evening consisted of some general discussion as to what we wanted to accomplish that weekend, as well as what we all wanted to see happen with the association. Plaques were made commemorating the date and attested to the fact that the attendees had been at the organizational meeting of the North American Snow Plowers Association – the proposed name. Also in attendance at this dinner were Peggy Allin and Jennifer Fails, who served as SIMA's first executive director.

On Saturday, we convened in the basement of my home at 8 a.m. and spent the next 10 hours hammering out the mission statement, the initial set of bylaws and the association's directives. The association's original name was dis-

envisioned the executive director's position to be a part-time position. Jen Fails would become the first executive director of SIMA and Allin Companies funded her position. Unfortunately – at least for the Allin Companies – the executive director position quickly became a full-time job for Fails. She left the following winter and we selected Tammy Higham to fill the position. At the time, we determined the position had to be full-time and Tammy ended up staying with SIMA for nearly 10 years.

The initial board of directors also came from this group and included myself as board president, Kyle Hansen as first vice president, Marc Young as second vice president, Bill Milbier as

---

"As the person who gave our venture its initial credibility, we wondered **how we would fare without Phil's wisdom** and advice."

---

carded as being too limiting in its scope and the late Phil Christian suggested Snow & Ice Management Association. It stuck. That evening we cooked steaks on the grill on the deck of my house and discussed the fledgling association's structure and growth strategy.

Funding was discussed at length and we agreed to each donate $1,000. It was agreed that since these were donations with no guarantees that the association would grow and thrive, we would not ask for the money back if the endeavor failed. If it succeeded, this would be our membership fee and we would never have to pay dues again as Charter Founding Members.

Additionally, I offered to house – rent free – the association's offices in my building at Allin Companies. We

secretary, and Rick Kier as treasurer. However, no terms were set as we didn't know how long the association would exist!

Rich Redfern left the board almost immediately because he was just starting up his business and could not participate at the level we needed. Phil Christian opted not to be a board member because he believed it would compromise his landscape-consulting business. Instead, SIMA hired Phil as a consultant for $1, which Kyle Hansen donated out of his pocket on spot.

Sadly, Phil passed away suddenly in September 1997. As the person who gave our venture its initial credibility, we wondered how we would fare without Phil's wisdom and advice. I'm certain he would be tremendously proud

of just how well it has gone.

In 1999, SIMA was financially secure enough to take over half of the executive director's salary, as well as her health and pension benefits. In January 2000, SIMA took over all funding of its staffing requirements and moved its offices, thus sustaining its' expenses for the first time. By February 2000, SIMA became totally self-sufficient. This was a significant milestone to achieve within four years of that June meeting. Personally, the amount of time and money Allin Companies donated – in excess of $260,000 – to the cause was significant, but it has certainly proven to be a worthwhile investment in the industry.

The first SIMA Symposium in June 1998 was originally planned as a get together of snowplow contractors from around the country to have an intense three days of networking. If 50 people attended we felt we could call it a success. We selected the hotel and Peggy made all the arrangements. We often operated this way early on – I would dream it and Peggy and Tammy would make it happen.

Seventy-five people attended that first symposium. On the last day attendees asked where we were having next year's event. A second symposium had never crossed our minds because we weren't sure the first one would get off the ground. In a hastily arranged conference with the hotel sales staff, we booked the 1999 SIMA Symposium before we left the building.

The second symposium in Pittsburgh drew twice as many attendees as the first. At that point we knew we had something going. By the 2002 SIMA Symposium in St. Louis, the event had grown to 560 attendees and more than 62 exhibitors. At that show, the first edition of "Managing Snow & Ice" hit the shelves. We were fortunate to have it catch on and become a helpful educational resource for many snow and ice contractors.

September 2000 also saw the first stand-alone industry magazine for snow contractors. It was a personal dream come true, and one we had worked on for several years. GIE Media, in partnership with SIMA, published that first magazine.

Since that time, SIMA has taken the magazine elsewhere to be published and GIE Media has gone out on its own with *Snow Magazine*. As the leading entity in the snow industry, GIE Media has partnered with me in many endeavors. This is a partnership I cherish and embrace.

The board meeting at the June 2002 SIMA Symposium in St. Louis marked the end of my direct involvement with the association. The torch was officially passed to others to guide and lead the organization going forward. At the time, I was convinced that this was best for SIMA. I still believe it was necessary so that the association did not stagnate under the same leadership. Some change is always good.

SIMA had grown way beyond what we envisioned at that first meeting in my basement back in 1996. None of us there envisioned an organization whose membership would grow to more than 1,000 members. None of us realized the impact that our weekend meeting would have on the industry.

The association is here to advance the industry. However, industry professionals are not always given the respect they deserve as businesspeople. SIMA grew steadily over its first decade and has since leveled out in membership. However, those who were there at the beginning found tremendous advantage to being members. There is strength in numbers. As an association, SIMA represents a significant segment of the industry. I am proud of what we accomplished.

SIMA has far exceeded our expectations and we all pray for the association's continued success. Those who are leading SIMA into the future are intelligent, well-respected and dedicated to the ideals that are at the association's core. I wish them luck and I have the utmost confidence in their ability to do what is right.

A Gold Medal Experience

In 2001, I quoted the snow removal contract for the 2002 Winter Olympic Games in Salt Lake City, Utah. We were awarded the entire contract for all 14 venues, spread over an area 50 by 70 miles. My company became the only single-sourced service provider of those Olympic Games. It was a resounding success and we performed in an exemplary fashion. The letter of recommendation received from the Olympic Organizing Committee after completion of this project still gives me goose bumps when I read it. It was our crowning achievement.

I have many warm and fond memories of our Olympic Games experience. I was able to witness several behind-the-scenes moments. We had access to all communications surrounding the opening ceremonies. We heard about toilets backing up, POTUS (the President of the United States) entering the building, about steering fans away from certain parts of the venue and the like.

One week before the opening ceremony, we were tasked to plow almost all of downtown Salt Lake City due to some logistical concerns expressed by the organizing committee about others' ability to keep the city's streets cleared during the games. We had one week to bring in enough equipment, put a plan together and staff this "extra" to our contract. We succeeded, of course, but scrambled all the same.

We took over plowing various parts of the 14 venues that made up the games'

Working at the 2002 Winter Olympics was a once-in-a-lifetime experience for me and our company.

sites on Oct. 3, 2001 – four months prior to the Olympic Games. The area received no snow in October and most of November. The first snowfall arrived Thanksgiving weekend and proceeded to dump 9 feet of snow at the Snow Basin Resort. We freaked out. We mobilized everyone we could find and kept the site clear, but it was quite a wake-up call. October and November were record-low snowfalls for the entire area, and December saw record-high snowfalls. Fortunately, it only snowed twice during the actual Olympic Games.

Our contract provided for liquidated damages for failure to perform. Thankfully, we never saw that paragraph implemented. The fines were exorbitant – $25,000 for each failure prior to the actual games and $50,000 for each failure once the games were taking place.

The closest we came was on one occasion when our staff was at Snow Basin – where the downhill skiing competition was held until 5 a.m. and then left because there was no snow falling and none in the forecast. However, at 6 a.m. it began snowing. By 6:15 a.m., there was an inch of snow and by 6:30 a.m. there were 2 inches of snow on the ground. By 6:45 a.m., our crew was back at the site working.

That afternoon I was called downtown for a meeting about what had happened. I went in armed to the teeth with reports, data, time sheets, etc., proving we reacted faster than any normal person could in that situation. My boss was understanding. However, his boss was not. When I handed her the stack of data she threw it up in the air and said, "I don't care. Your contract states at 2 inches of snow you should be out there working and you were 15 minutes late."

We were 15 minutes late. I was told that if it happened again it would cost us $25,000. The meeting was short and to the point. When I walked out of the building I told my general manager we would *not* go through that again. From that day forward, that site was staffed 24/7 just in case. We never got that close to failure again.

Riding The Business Rollercoaster

After the Olympic Games, my company grew by leaps and bounds. Eventually, we were managing snow in 42 states, overseeing revenues and expenditures in excess of $40 million. Recognized as the first national snow removal company, we became victims of our own success.

Not paying heed to advice I normally hand out to others, we did not operate "by the numbers" and eventually ran into cash-flow issues, which ultimately led me to sell my company. The people who bought the company floundered under the weight of their own debt and eventually failed, This resulted in numerous contractors being "stung" for over $9 million. Many blamed me for this, even though I was long gone from the company. However painful as it was, that is the ugly side of the business world that we all have endured at one time or another.

As a result of the Olympics experience, I invented a portable machine that melted snow. I was fortunate to partner with a public entity that wanted to finance the startup of Snow Dragon. I had learned owning 20% of something well capitalized was much better than owning 100% of something that always seemed cash starved.

With this venture, I was instructed to "take it international" and I did. I visited 36 countries in 36 months – some two or three times – to set up distribution channels around the world. Our biggest customer was the City of Moscow. Snowmelting came to the forefront of the snow contracting world through hard work and dedication of a staff bent on success. Eventually, I sold my interest in the venture and left the company in 2009.

Stepping Out On My Own (Again)

Encouraged by my wife and this book's publisher, I began a consulting practice dedicated to working with snow contractors. I've said many times that my successes are more a result of an intense fear of failure than any one indicator of success. While reluctant to expose myself to the risk of failure, I was amazed by how the industry accepted my consulting practice. I regularly write about the snow industry as I work with contractors to increase their revenues and profits.

When those who seek my advice and guidance ask what I bring to the table, I believe I offer an honest response. There are numerous land mines out there that any business person can step on and have their entire business blow up on them. I have stepped on just about every land mine a snow contractor can step on, and I have lived through it all to tell others how, why, where and when it happens.

I am personally grateful that eight people who knew very little about me came to Erie, Pa., in 1996, not knowing what to expect. They came on faith in an idea that was little more than a dream, and placed trust in someone they really didn't know to do what was right for the industry. For that faith, I'd like to express my sincerest thanks.

The snow industry has been very good to me and I love it intensely. I hope to be a part of it for a very long time.

The hours were long and the demands many during our 2002 Olympic Games experience.

INDEX

PHOTO CREDITS

Cover Image
Istockphoto..

About the Author
Paul Lorei 7

Chapter One
Donn Angus.. 14
National Oceanic & Atmospheric Administration 17

Chapter two
Dreamstime/shado59 20
Fisher Engineering... 24

Chapter 3
Thinkstock/stockbyte.. 31
Gilles Delisle... 33
Stockbyte.. 36

Chapter 4
Illustration by Mike Grygo 39
Illustration by Mike Grygo 41
Illustration by Mike Grygo 42
Thinkstock/Jupiterimages 45
Illustration by Neave Landscaping.. 48

Chapter 5
Mike Grygo 52
Donn Angus... 57

Chapter 6
Dreamstime/Douglas Hockman 61
Thinkstock/Gettyimages.. 63
Thinkstock 64
Photodisc.. 67

Chapter 7
Thinkstock 69
Shutterstock/Ron Hilton 70
Istockphoto/Skhoward... 72
weather.com 74
Thinkstock 76